AF445473

Numerical and Experimental Analysis of the Fracture Behaviour of Heterogeneous Welded Structures

Numerical and Experimental Analysis of the Fracture Behaviour of Heterogeneous Welded Structures

Editors

Dražan Kozak
Nenad Gubeljak
Aleksandar Sedmak

MDPI • Basel • Beijing • Wuhan • Barcelona • Belgrade • Manchester • Tokyo • Cluj • Tianjin

Editors

Dražan Kozak
Mechanical Engineering Faculty
in Slavonski Brod
University of Slavonski Brod
Slavonski Brod
Croatia

Nenad Gubeljak
Faculty of Mechanical
Engineering
University of Maribor
Maribor
Slovenia

Aleksandar Sedmak
Faculty of Mechanical
Engineering
University of Belgrade
Belgrade
Serbia

Editorial Office
MDPI
St. Alban-Anlage 66
4052 Basel, Switzerland

This is a reprint of articles from the Special Issue published online in the open access journal *Materials* (ISSN 1996-1944) (available at: www.mdpi.com/journal/materials/special_issues/ Analysis_Fracture_Heterogeneous_Welded_Structures).

For citation purposes, cite each article independently as indicated on the article page online and as indicated below:

LastName, A.A.; LastName, B.B.; LastName, C.C. Article Title. *Journal Name* **Year**, *Volume Number*, Page Range.

ISBN 978-3-0365-3462-6 (Hbk)
ISBN 978-3-0365-3461-9 (PDF)

Contents

About the Editors

Dražan Kozak

Dražan Kozak is a Full Professor at the Mechanical Engineering Faculty, University of Slavonski Brod, Croatia. He is also a Rector's Assistant for quality assurance at the same university. He was a Vice-Rector for education and students at the Josip Juraj Strossmayer University of Osijek, Croatia (2017–2020), Dean of the Mechanical Engineering Faculty in Slavonski Brod (2009–2013), and Vice-Dean for science (2005–2009) at the same institution.

He has achieved an M.Sc. degree in Mechanical Engineering (1995) and a Ph.D. degree (2001) in Construction Design from the Faculty of Mechanical Engineering and Naval Architecture, University of Zagreb. He teaches courses on mechanics, the strength of materials, numerical methods, finite element method, structural optimization, fracture mechanics, experimental mechanics, etc., at the University of Slavonski Brod, Croatia.

His research interests include structural integrity assessments of heterogeneous welded joints, structural health monitoring, the endurance of 3D-printed implants made from Ti alloys, mechanical properties of porous lattice structures, etc. As the main researcher, he has led 5 bilateral international projects, 8 CEEPUS projects, several EU projects, and many national projects granted by Croatian Science Foundation. He has published 113 articles in impactful journals referenced in the Web of Science databases, which have been cited 691 times. His h-index is 14 according to WoSCC, and 16 according to the Scopus database. He is an Honorary Professor at the GAMF Kecskemet college, Hungary, and Guest Professor at the Faculty of Technical Sciences, University of Novi Sad, Serbia. Prof. Kozak is a member of the Editorial Boards of 14 journals. He was the President of TEAM International Society, supervisor of 8 defended dissertations, and member of the board for assessment and defense of the doctoral dissertations for 24 candidates in Croatia and abroad. He has authored 7 university books, and is the President of the ESIS Croatia.

Nenad Gubeljak

Full Professor Nenad Gubeljak studied Mechanical Engineering at the University of Maribor in Slovenia, where he completed his M.Sc. in 1991 and Ph.D. in 1998. During his postgraduate study, he was awarded an Arge Alpe-Adria scholarship in Austria, a Slovenian national foundation grant, and a DAAD scholarship in Germany. For his contribution to the behaviour of mismatch welded joints, he received the Henry Granjon prize from the International Institute of Welding (IIW) in the "Integrity of welded constructions"category in 1999. For the project "Mis-matching effect of constraint", he earned a Marie Skłodowska Curie post-doctoral scholarship from the European Commission, and was a guest scientist at the GKSS research centre Geesthacht in Germany in 2000–2001. As a member of a consortium from Slovenia in the period from 2003 to 2012, he led EU projects in the framework FP6, FP7 (Fitnet, Hiperc), as well as CemLib in Eureka projects MOSTIS and OLMOST. In 2011, he was named Guest Professor at the Institute of Mechanics, Universite de Lille, France. In the last 10 years, he has led bilateral projects with foreign science institutions and universities in Argentina, Belgium, the United States, France, Austria, Czechia, Russia, Croatia, and Serbia. As a member of a consortium from Slovenia, he leads the KOMET 2 project of the Austrian Academy of Sciences. He is a member and a representative for Slovenia in the European Association for the Integrity of Constructions ESIS. He is a member of the American Association for Standardisation of Material Testing, ASTM section E0.8 for fractures and fatigue. At the Faculty for Mechanical Engineering in Maribor, he is head of the Chair for Mechanics. Mostly in cooperation with foreign scientists, he has published more than 180 scientific research articles in international journals, as well as contributions in international conferences. These papers are related to fatigue and the fracture behaviour of structural materials and welded joints.

Aleksandar Sedmak

Aleksandar Sedmak is Professor Emeritus at the Faculty of Mechanical Engineering, University of Belgrade, Serbia, a Visiting Professor at Drexel University, USA, (1999–2002), Assistant Minister for Science and Technology Development in the Serbian government (2003–2006), Vice-Rector for International Cooperation, University of Belgrade, (2006–2009), and Director of the Innovation Center of the Faculty of Mechanical Engineering in Belgrade, (2006–2021). He has been teaching since 1979 at the Faculty of Mechanical Engineering, University of Belgrade, mainly in engineering materials, welding, numerical methods, fracture mechanics, and structural integrity. He has been the advisor for 60 D.Sc. theses. He has published 138 papers in WoS journals and 319 papers in the Scopus database, with 2046 citations (h=21).

Prof. Sedmak is the President of Serbian Structural Integrity and Life Society and Editor-in-Chief of the *Structural Integrity and Life* journal (Scopus, eSCI), the Vice-President of the European Structural Integrity Society (ESIS), a member of the Editorial Boards and reviewer for a number of prominent global journals in fracture mechanics, and was the Chairman of the European Conference on Fracture (ECF22), held in Belgrade in 2018. He has been the Guest Editor of Special Issues of *Engineering Failure Analysis*, *Engineering Fracture Mechanics*, *Theoretical and Applied Fracture Mechanics* and the *International Journal of Fatigue*.

Prof. Sedmak is a longstanding member of prominent societies, such as ESIS, AWS and ASME, as well as an Honorary professor at Brasov University, Romania, since 2016, and foreign member of the Hungarian Academy of Engineering Sciences since 2017, a member of the Serbian Academy of Engineering Sciences since 2012, and an Honorary Member of the Italian Group on Fracture (IGF) since 2020.

Preface to "Numerical and Experimental Analysis of the Fracture Behaviour of Heterogeneous Welded Structures"

Welding is the most widespread technology for the connection of different materials, elements and structures. Loading capacity and knowledge about stress–strain behaviour is most important for safety and the reliable use of structures for different purposes in energy supplies and the transport of good and people. Filler material used in welding should be the same class as the base material; however, welding codex often uses a filler with a higher yield strength than the base material (such as by repair welding). Welded joints are also often the location of potential flaws, where flaw assessments assume homogeneous material properties, although welds are heterogeneous. There is a compendium of yield load solutions for mismatched strength fracture mechanics specimens developed to address the heterogeneity in the weld joint. However, solutions for the yield load at different combinations of the strength mismatch within the weld are missing, where mechanical testing and finite element simulations are necessary. In addition to more conventional approaches, a multi-scale approach recently introduced in the assessment of weld heterogeneity sounds very promising. It is also efficient to consider residual stresses, which can strongly affect the stress distribution around flaws in heterogeneous weldments. The multi-scale methodology is computationally efficient and provides a possible means to bridge multiple length scales (from 10 nm in MD simulation to 10 mm in FE models). This could be a useful tool by considering an acceptable level of accuracy with respect to yield load in heterogeneous welds.

In this book, modern trends in testing and simulating heterogeneous welded joints, including multi-scale approaches, resulting in appropriate flaw assessment procedures, are highlighted and discussed. The eleven research papers presented in this book give some overview of recent analytical, numerical, and experimental investigations in the field of yield strength mismatched welded joint behaviour. The papers cover several important issues to more accurately characterise the fracture mechanics behaviour and structural integrity assessment, as follows:

- New mathematical model for the determination of yield loads for the present yield strength mismatch in weld configuration at different crack positions;
- New methodology for determining the actual stress–strain diagram based on analytical equations, in combination with numerical and experimental data, using 3D digital image correlation (DIC);
- Simulation of local variation in material properties of the fracture behaviour in a multi-pass mismatched X-weld joint;
- Experimental procedures include the characterization of average material properties by tensile testing and evaluations of base and weld metal resistance to stable tearing by the fracture testing of fracture mechanics specimens containing a weld notch;
- Behaviour of welded joints in the simultaneous presence of several different types of multiple defects, such as linear misalignments, undercuts, incomplete root penetration and excess weld metal;
- Modified equation for estimating the C^* integral for a welded compact tension (CT) specimen under creep conditions, etc.

The editors of this book hope that the readers will find this chapters interesting and useful for their everyday work in this challenging area.

Dražan Kozak, Nenad Gubeljak, Aleksandar Sedmak
Editors

materials

MDPI

Article

Yield Load Solutions for SE(B) Fracture Toughness Specimen with I-Shaped Heterogeneous Weld

Pejo Konjatić [1,*], Marko Katinić [1], Dražan Kozak [1] and Nenad Gubeljak [2]

[1] Mechanical Engineering Faculty, University of Slavonski Brod, Trg Ivane Brlic Mazuranic 2, 35000 Slavonski Brod, Croatia; mkatinic@unisb.hr (M.K.); dkozak@unisb.hr (D.K.)

[2] Faculty of Mechanical Engineering, University of Maribor, Smetanova 17, SI-2000 Maribor, Slovenia; nenad.gubeljak@um.si

* Correspondence: pkonjatic@unisb.hr

Abstract: The objective of this work was to investigate the fracture behavior of a heterogeneous I-shaped welded joint in the context of yield load solutions. The weld was divided into two equal parts, using the metal with the higher yield strength and the metal with the lower yield strength compared to base metal. For both configurations of the I-shaped weld, one with a crack in strength in the over-matched part of the weld and one for a crack in the under-matched part of the weld, a systematic study of fracture toughness SE(B) specimen was carried out in which the crack length, the width of the weld and the strength mismatch factor for both weld metals were varied, and the yield loads were determined. As a result of the study, two mathematical models for determination of yield loads are proposed. Both models were experimentally tested with one strength mismatch configuration, and the results showed good agreement and sufficiently conservative results compared to the experimental results.

Keywords: yield load; heterogeneous weld; numerical analysis; SE(B) specimen

Citation: Konjatić, P.; Katinić, M.; Kozak, D.; Gubeljak, N. Yield Load Solutions for SE(B) Fracture Toughness Specimen with I-Shaped Heterogeneous Weld. *Materials* **2022**, *15*, 214. https://doi.org/10.3390/ma15010214

Academic Editor: Ruitao Qu

Received: 23 November 2021
Accepted: 25 December 2021
Published: 28 December 2021

Publisher's Note: MDPI stays neutral with regard to jurisdictional claims in published maps and institutional affiliations.

1. Introduction

Joining metals by welding is nowadays widely used in the construction of most engineering structures. The requirements for high quality welded joints joining similar or dissimilar metals, taking into account the mechanical properties of the metal, lead to the production of welded joints with significant differences in strength compared to the base metal.

Like all structures, welded structures are susceptible to damage during use, particularly in the weld or heat-affected zone, due to the change in metal properties and the expected significant nonlinear deformations caused by mechanical heterogeneity. Repaired welds are commonly used in steel structures either to correct initial fabrication defects or to repair damage during service to extend the service life of the structure [1]. When welds are repaired, additional heterogeneity is introduced into the already heterogeneous structure.

In the conventional evaluation of the safe operation of defect-free structures, the applied stresses are compared to a limit stress, such as the yield strength of the material. When damage in the form of a crack is present, the assessment of welded joints is based on the evaluation of the stress intensity factor, the J-integral and the crack tip opening displacement (CTOD) [2]. On the other hand, the influence of mechanical heterogeneity on the fracture behavior of welds is not explicitly included in the mentioned fracture mechanics parameters. However, methods and procedures for evaluating homogeneous and heterogeneous structures, which have been developed recently, can be used to determine whether or not the structure is safe for further exploitation.

One commonly used procedure for a structural integrity assessment is the SINTAP procedure (<u>S</u>tructural <u>INT</u>egrity <u>A</u>ssessment <u>P</u>rocedure) [3]. The application of the SINTAP procedure is based on the implementation of the yield load solution in the failure assessment diagram (FAD) to determine the safe operation of the assessed structure. There are a number of studies dealing with various aspects of the fracture behavior of homogeneous welds with strength mismatch compared to the base metal, including recent ones [4–9], as well as a number of studies dealing with heterogeneity in welds with strength mismatch [10–16].

A common parameter for describing the level of strength mismatch between individual metals, in the context of this investigation, between base metal and weld metals, i.e., mismatch in yield strength between the weld metal and the base metal, is quantified by the mismatch factor M:

$$M = \frac{\sigma_{YW}}{\sigma_{YB}} \tag{1}$$

where σ_{YW} and σ_{YB} represent the yield strength of the weld metal and the yield strength of the base metal (BM), respectively, while $M < 1$ refers to under-matching (UM) and $M > 1$ to over-matching (OM).

Yield load solutions are available for a limited number of strength mismatch configurations for over-matched and under-matched welds [17–20], but only very limited and partial solutions in situations where additional heterogeneity due to repair weld metal is present in another level of the strength mismatch [21,22]. Therefore, this research aims to extend the existing yield load solutions to I-shaped heterogeneous weld solutions in order to gain insight into the fracture behavior of the repaired weld and open the possibility of applying structural assessment procedures for repair welds. As a result, a compendium of yield load solutions for a standard fracture mechanics specimen SE(B) with a heterogeneous weld is given, which can be used as the input parameter for an assessment using standard structural integrity assessment procedures.

2. Problem Description and Investigation Plan

Since butt welds are used extensively in the welding industry, there is often a need to repair such welds when defects occur during welding or during the service life of the welded structure. If the repair involves the use of a filler metal different from the filler metal used to weld the original weld, the result is a heterogeneous welded joint with two different weld metals in addition to the base metal. When structures are put back into service after repair, the occurrence of cracks in the original part or in the repaired part of the weld is possible again.

For this reason, and for the reasons given in the introduction, a study of the fracture behavior of an I-shaped butt weld was carried out. The effects of weld damage in the form of a crack were analyzed.

Due to the complexity of the problem to be analyzed, it was necessary to introduce certain idealizations and simplifications. In all previous studies on a similar topic, several such idealizations were introduced, starting from the weld geometry idealized by a rectangle, and the observed cracks were located at the interface of dissimilar materials or in the middle of the weld due to the nature of crack formation described in [18,23,24]. In this study, the I-shaped weld was also idealized as a rectangular shape, as well as the original and repaired part of the weld (Figure 1), and the crack was located in the center of the weld. In [25], researchers have demonstrated that the mechanical properties of the heat-affected zone have a negligible effect on the stress concentration at the crack tip when the crack tip is located in the center of the weld. However, if the crack tip is located in or near the heat-affected zone, the properties of the heat-affected zone have to be taken necessarily into account [26]. Since the crack in the middle of the weld was analyzed here, the heat-affected zone was omitted.

The difference in elastic properties of the material as well as the strain hardening of the material affect the fracture behavior of the weld, but here, only the influence of the degree of strength mismatch between the single welded metal and the base metal is studied. In addition to the strength mismatch, the change in weld width and the crack size were also analyzed. The influence of mentioned geometrical and mechanical parameters on the yield load was observed, i.e., the load at which the metal flows through the entire cross-section of the weld, since at that moment a plastic hinge is formed.

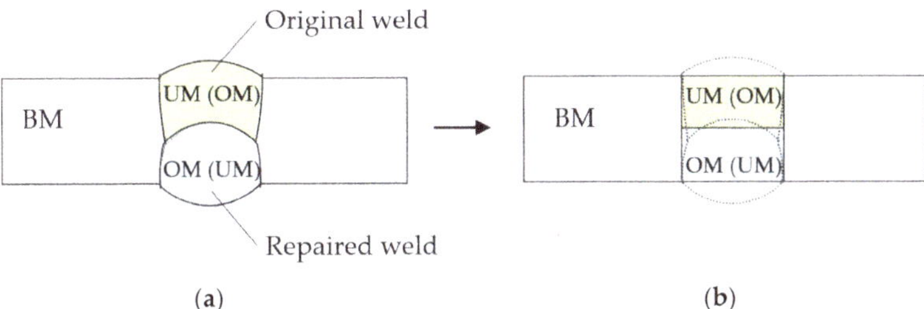

Figure 1. Idealization of repaired I-shaped butt-welded joint: (**a**) heterogeneous welded joint; (**b**) idealized heterogeneous welded joint.

3. Finite Element Analysis

In order to investigate the influence of weld material heterogeneity on weld fracture behavior, the weld area was divided into two zones of equal size but consisted of different metals. The first zone represented the original weld before repair, while the second zone represented the repaired portion of the weld. In the first variant, the zone of the original weld was made of a metal whose yield strength was lower than the yield strength of the base metal (UM), while the second half of the weld was made of a metal whose yield strength was higher than the yield strength of the base metal (OM). In the second variant, positions of the UM and OM part of the weld were reversed. Combinations where both weld metals have over-match or under-match character were not covered by this investigation.

Due to the possibility of crack formation in the original and the repaired part of the weld, both variants were analyzed. Due to geometry and load symmetry, a plane strain two-dimensional numerical model of one half of an SE(B) specimen with homogeneous weld metal (WM) was created in ANSYS [27] (Figure 2a,c). The model was verified comparing finite element results with the analytical method of slip line field analysis [17] that is used in the analytical analysis of strength mismatch welds. Results of verification showed very good agreement between the results of numerical and slip line field analyses, and this verification is already published in [21]. A single change was made to the verified numerical model, in the form of splitting the homogeneous weld into two equal portions of over-matched and under-matched weld metal to form a heterogeneous weld (Figure 2b,c).

To determine the influence of weld width H and crack length a on the yield load, the width of the weld H was varied as $H = W/2$, $H = W/4$, $H = W/8$, $H = W/16$ and $H = W/24$, while the crack length in relation to the height of the specimen W was varied as $a/W = 0.1$, $a/W = 0.2$, $a/W = 0.3$, $a/W = 0.4$ and $a/W = 0.5$ (Figure 2d). The length of the specimen S was kept constant.

The base metal (BM) and the weld metals (OM and UM) were modeled as isotropic linearly elastic and nearly ideally plastic materials with Poisson's ratio of 0.3 and with a Young's modulus of 202 GPa for base metal, 200 GPa for over-matched and 206 GPa for under-matched metal. Elasticity mismatch also have an influence on the fracture behavior of a welded joint [28,29], but this slight degree of elasticity mismatch did not show an influence on the values of the obtained yield loads compared to ones obtained without elasticity mismatch. The yield strength of the base metal was 545 MPa. The strength mismatch of base and weld metals are varied on three levels: over-match metal with mismatch factor $M_{OM} = 2$, 1.5 and 1.19 and under-match metal with mismatch factor $M_{UM} = 0.86$, 0.75 and 0.5. Yield strength and mismatch factors $M_{UM} = 0.86$ and $M_{OM} = 1.19$ were chosen due to later comparison to experimental results.

Due to the faster convergence of the results, a practically negligible strain hardening exponent was used, which did not affect the results but significantly reduced the computation time. To avoid the incompressibility problem, an isoparametric planar element with eight nodes, plane strain and reduced integration was used. Singular elements with a size of 100 μm were used in the first ring of elements around the crack tip to produce the square root singularity of the stress–strain field. Models were meshed with 1847 finite elements and with 5690 nodes. Prepared models were loaded with a load large enough to cause the material to yield through the entire cross-section of the model.

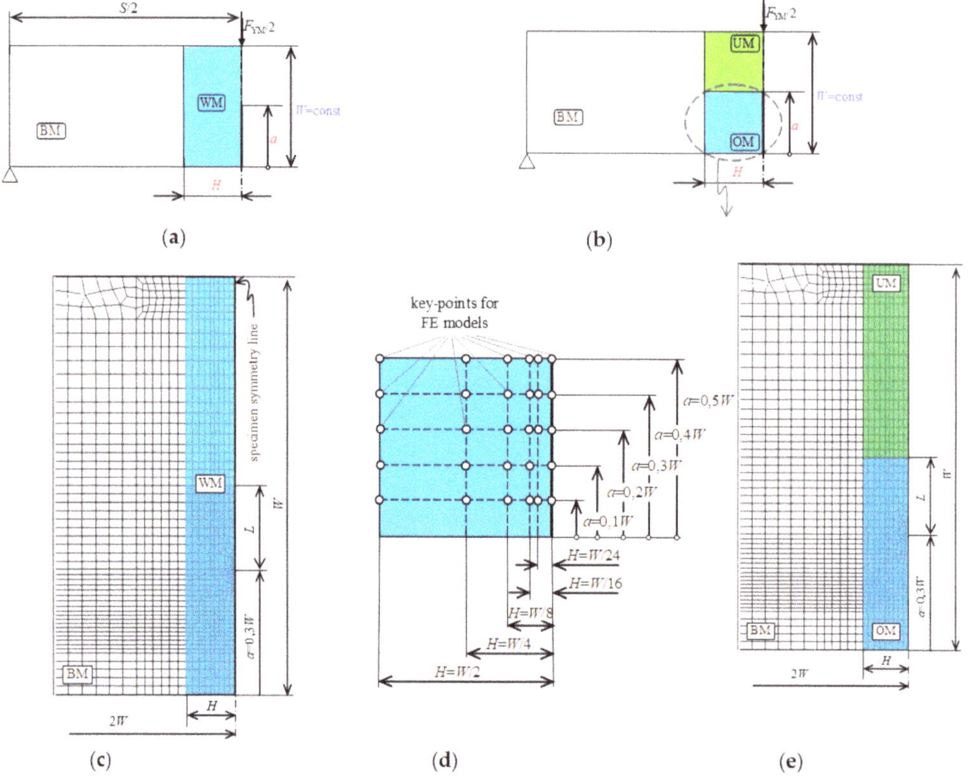

Figure 2. Numerical model: (**a**) homogeneous weld for verification; (**b**) heterogeneous weld; (**c**) detail of finite element mesh of homogeneous weld for verification; (**d**) key-points for variation of weld width H and crack length a; (**e**) detail of finite element mesh of model with heterogeneous weld.

The load was increased gradually in small increments to accurately determine the load of plasticization of the entire net section of the specimen, indicating the formation of a plastic hinge and plastic collapse. As a criterion for material flow, the von Mises criterion was used. A total of 450 simulations were performed for a crack located in an over-matched and under-matched part of the weld.

4. Results of Finite Element Analysis

Obtained yield loads for heterogeneous weld were normalized with yield loads of all base specimen according to [17] and presented in diagrams depending on the weld slenderness $(W - a)/H$. To facilitate the interpretation of the results, the dependence of weld slenderness on crack length and weld width is shown in Figure 3. It can be seen that the slenderness of the weld increases significantly with decreasing weld width and becomes less pronounced with decreasing crack length.

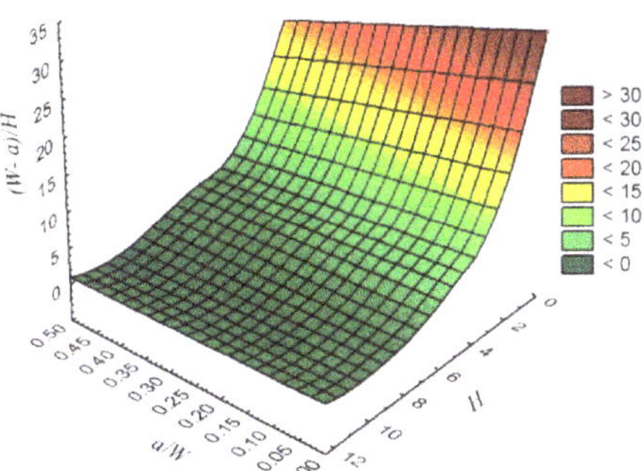

Figure 3. Dependence of weld slenderness $(W-a)/H$ on weld width H and crack length a/W.

4.1. Yield Load Solutions for a Crack in the Over-Matched Part of the Weld

Yield load solutions as a result of the analysis of a heterogeneous weld with a crack in the over-matched part of the weld in the function of weld slenderness $(W-a)/H$ are presented in Figure 4.

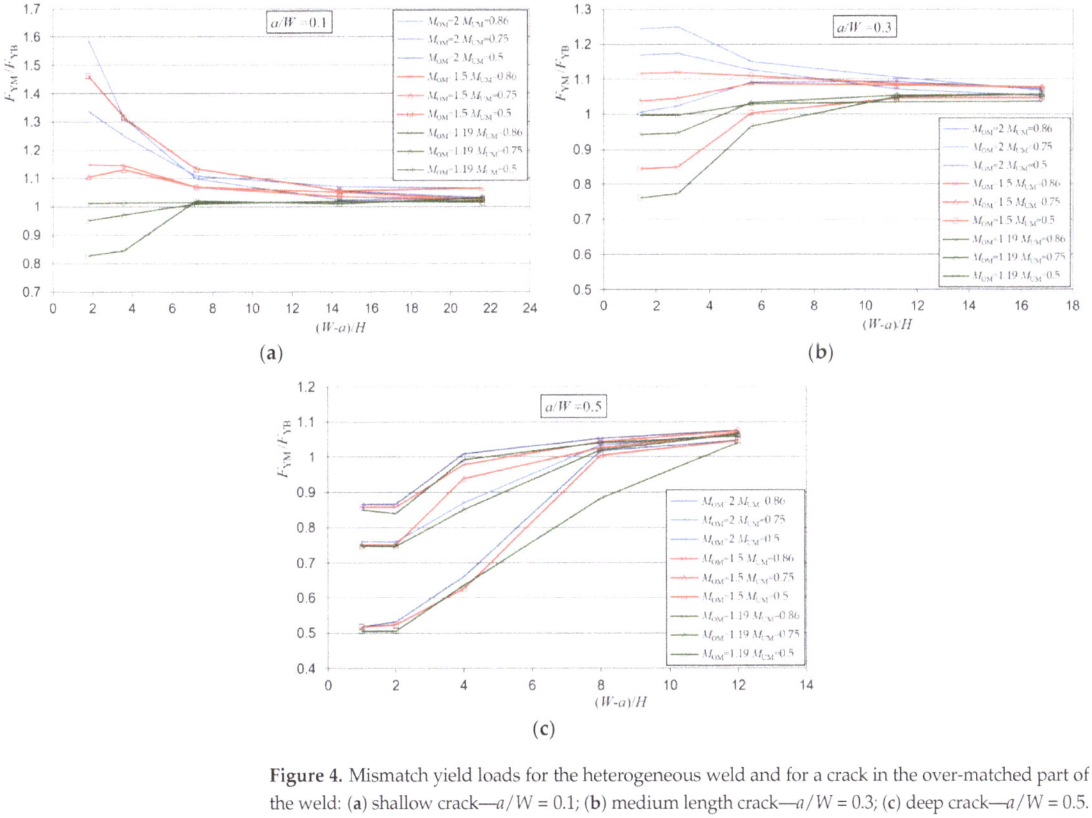

Figure 4. Mismatch yield loads for the heterogeneous weld and for a crack in the over-matched part of the weld: (**a**) shallow crack—$a/W = 0.1$; (**b**) medium length crack—$a/W = 0.3$; (**c**) deep crack—$a/W = 0.5$.

From Figure 4, it can be seen that the dispersion of the yield load solutions at lower weld slenderness depends on the present weld metals, while at higher weld slenderness, the solutions of all metal combinations and all crack lengths approach an asymptotic value. This value is slightly higher than the value 1, indicating a slight increase in the strength of the weld compared to the component of the homogeneous base metal.

When the slenderness of the weld is lower, different effects occur depending on the length of the crack in the weld. For the crack $a/W = 0.5$, there is only under-matched metal in front of the crack, which is represented by the mismatch factor M_{UM}. Therefore, the solutions were the values 0.5, 0.75 and 0.86 because the dominant metal is in front of the crack. Although the solutions were these values, it is noticeable that they are actually slightly larger, which is a consequence of the formation of the yield zone partially through the over-matched metal too, which has a higher value of the mismatch factor M_{OM}.

Figure 5 shows formation of the yield zone in a heterogeneous weld with a crack in the over-matched part of the weld for varying weld width H and constant crack length $a/W = 0.5$ for mismatch factors $M_{OM} = 1.19$ and $M_{UM} = 0.86$.

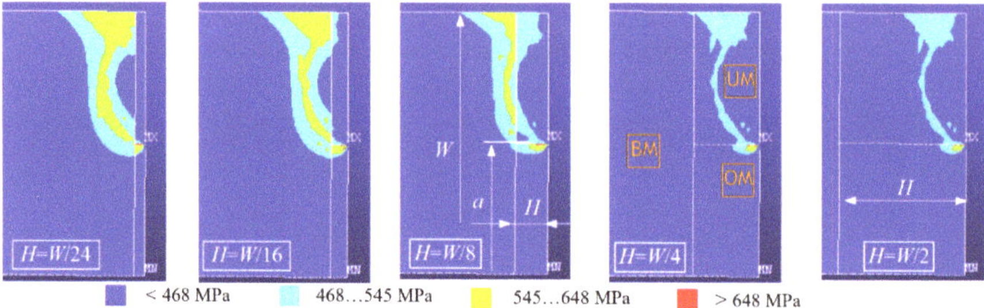

Figure 5. Formation of the yield zone in a heterogeneous weld with a crack in the over-matched part of the weld for varying weld width H and constant crack length $a/W = 0.5$.

The appearance of the yield zones is similar for all geometries, and depending on the width of the weld, the yield zone extends through two or all three materials. For narrow weld widths $H = W/24$, $H = W/16$ and $H = W/8$, the material yield zone spreads through the base metal and both weld metals, while for wider welds $H = W/4$ and $H = W/2$, the yield zone stays within the weld metal.

As the size of the crack decreases, the metal in which the crack is located becomes more influential. This is particularly pronounced for combinations of metals whose mismatch factors M_{OM} and M_{UM} differ significantly, while the solutions for combinations of metals with closer values of M_{OM} and M_{UM} approach the values of 1 of the base metal. For example, for the combination of metals $M_{OM} = 2$ and $M_{UM} = 0.5$, the solutions range from 0.5 to 1.6, and for the combination of $M_{OM} = 1.19$ and $M_{UM} = 0.86$, the solutions are almost everywhere uniform and closer to the value 1.

The results of the numerical analyses for a crack in the over-matched part of the weld were processed in the software package TuringBot [30] using a symbolic regression algorithm to derive mathematical formulas from numerically obtained values with high efficiency. An equation that estimates the values of the ratio of the yield loads for the heterogeneous weld and the whole base metal was obtained. A high goodness-of-fit of the selected model was confirmed with the R-squared value 0.938 and RMS error 0.03748. The equation considers values of the over-match strength mismatch M_{OM}, under-match strength mismatch M_{UM}, the weld width H and the crack length a/W:

$$\frac{F_{YM}}{F_{YB}} = 1 - \frac{1 + H\left[\left(M_{UM} - M_{OM}\frac{a}{W}\right)\left(1 + 2\frac{a}{W} - \frac{1}{H}\right) + M_{OM} - 2\right]}{-20 - \left(\frac{H}{3} + 3\right)(H - 10)\frac{a}{W}} \tag{2}$$

4.2. Yield Load Solutions for a Crack in the Under-Matched Part of the Weld

Yield load solutions as a result of the analysis of a heterogeneous weld with a crack in the under-matched part of the weld in the function of weld slenderness $(W - a)/H$ are presented in Figure 6.

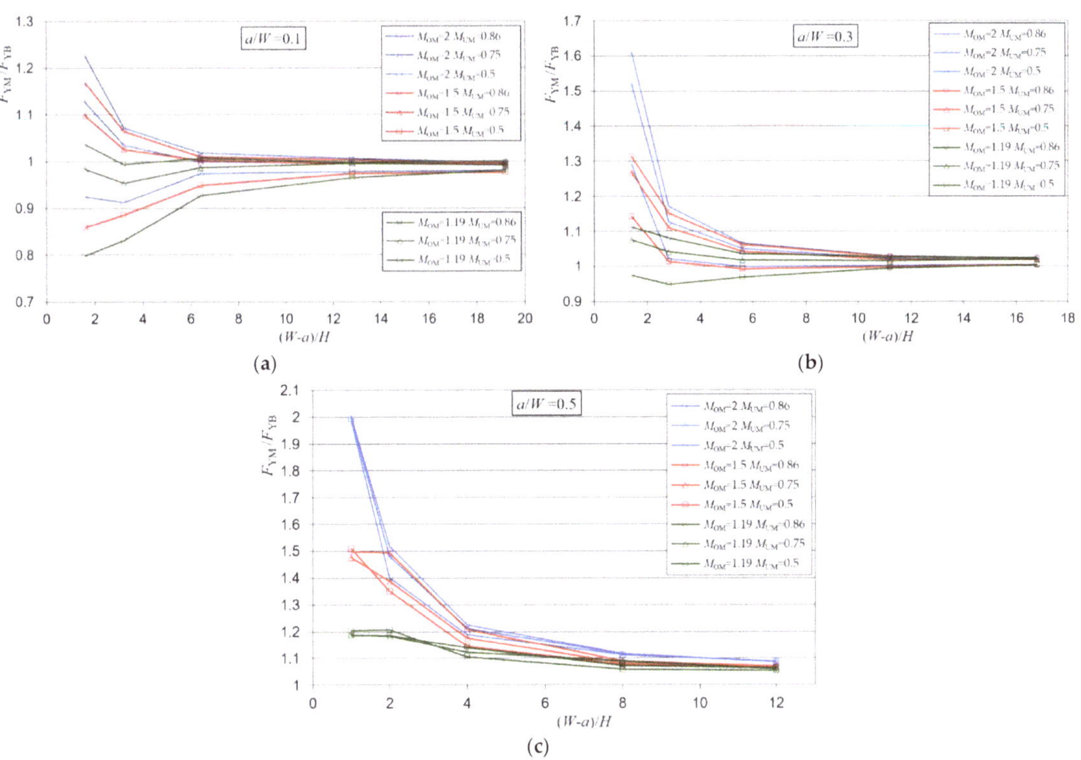

Figure 6. Mismatch yield loads for the heterogeneous weld and for a crack in the under-matched part of the weld: (**a**) shallow crack—$a/W = 0.1$; (**b**) medium length crack—$a/W = 0.3$; (**c**) deep crack—$a/W = 0.5$.

Results of the analysis, shown in Figure 6, indicated that for lower weld slenderness, the yield load solutions differ depending on the weld metals present in the weld, while for higher weld slenderness, the yield solutions of all metal combinations, as well as for all crack lengths, approached the value 1.

When the weld was less slender, the weld showed different behavior depending on the length of the crack in the weld. For the crack $a/W = 0.5$, only the OM was in front of the crack, therefore the yield loads were the values 1.19, 1.5 and 2. This happens because the metal in front of the crack, which has the over-match characteristic, takes the dominant role. Although the solutions were these values, it can be noted that they were somewhat lower, which was a consequence of the partial propagation of the yield zone also through the under-matched metal.

As the length of the crack decreases, the metal in which the crack is located also becomes more influential. Similar to the case where the crack was in an over-matched metal, it can be observed that the solutions with closer values of M_{OM} and M_{UM} approached the values of 1 of the base metal. For example, for the combination of $M_{OM} = 2$ and $M_{UM} = 0.5$, the solutions ranged from 0.9 to 2, and for the combination of $M_{OM} = 1.19$ and $M_{UM} = 0.86$, the solutions were almost uniform and were everywhere closer to the value of 1.

The results of the analysis for a crack in the under-matched part of the weld were also processed in the software package TuringBot using a symbolic regression algorithm and an equation for the estimation of the values of the ratio of yield loads for the heterogeneous weld and the whole base metal was obtained:

$$\frac{F_{YM}}{F_{YB}} = \frac{H - \left[H \cdot M_{OM}\left(M_{UM} + \frac{a}{W}\right) - \left(M_{OM} - 3\frac{a}{W}\right)\left(M_{OM} - \frac{1}{H}\right) \right]}{\frac{23}{M_{UM}} \frac{a}{W} - 34} + 1 \tag{3}$$

A high goodness-of-fit for model with a crack in the under-matched part of the weld was confirmed with the R-squared value 0.953 and RMS error 0.04095.

5. Experimental Investigation

For this investigation, standard SE(B) test specimens were prepared from the welded plate. For the base metal (BM), NIOMOL 490 was used as a high-strength, low-alloy, fine-grain steel in the hardened and tempered condition according to the HT 50 grade. Using the flux cord arc welding procedure and two tubular wires as filler material FILTUB 75 and VAC 60 as an over-match and under-match material, a heterogeneous weld was produced with the strength mismatch factor 1.19 and 0.86. Mechanical properties of the base metal and OM and UM part of the weld, shown in Table 1, were obtained by a tensile test. Five round specimens with a 5 mm diameter were used for each metal. The position and orientation of round specimens in the weld joint are shown in Figure 7a. The chemical composition of BM, UM and OM metal, shown in Table 2, is provided by the manufacturer, where OM and UM chemical composition is provided for pure weld metal.

Table 1. Mechanical properties of base and weld metals with mismatch factor.

Material	$R_{p0.2}$, MPa	R_m, MPa	E, GPa	M
Base metal (NIOMOL 490)	545	648	202	-
Over-matched (FILTUB 75)	648	744	184	1.19
Under-matched (VAC 60)	468	590	206	0.86

Table 2. Chemical composition of base and weld metals.

Material	C	Si	Mn	P	S	Cr	Mo	Ni
Base metal (NIOMOL 490)	0.123	0.33	0.56	0.003	0.002	0.57	0.34	0.13
Over-matched (FILTUB 75)	0.040	0.16	0.95	0.011	0.021	0.49	0.42	2.06
Under-matched (VAC 60)	0.096	0.58	1.24	0.013	0.160	0.07	0.02	0.03

For fracture toughness testing, specimens were prepared, and a single-sample method was used according to the standard BS 7448 [31]. CTOD fracture toughness specimens with dimensions and notch orientation are shown on Figure 7. The CTOD tests were carried out at room temperature (+24 °C) under displacement control (1 mm/min). Load F, total displacement, crack tip (CTOD) and crack mouth opening displacement (CMOD) were recorded during the tests. Tests were performed for two configurations: with a crack in the over-matched part of the weld and with a crack in the under-matched part of the weld. A total of 14 specimens were tested: 7 specimens with a crack initiated in the OM part of the weld and 7 with a crack initiated in the UM part of the weld.

The plots of load versus CMOD were obtained and shown in Figure 8a. During fatigue pre-cracking in two specimens, with a notch in the OM part of the weld, a crack reached the fusion line between the OM and UM and advanced to the UM part of the weld (Figure 8a shown with dotted lines); therefore, they are omitted in later comparison with the yield load solutions for a heterogeneous weld. For every sample, a crack location (OM or UM) and initial crack length compared to the height of the specimen (a/W) is shown in the legend.

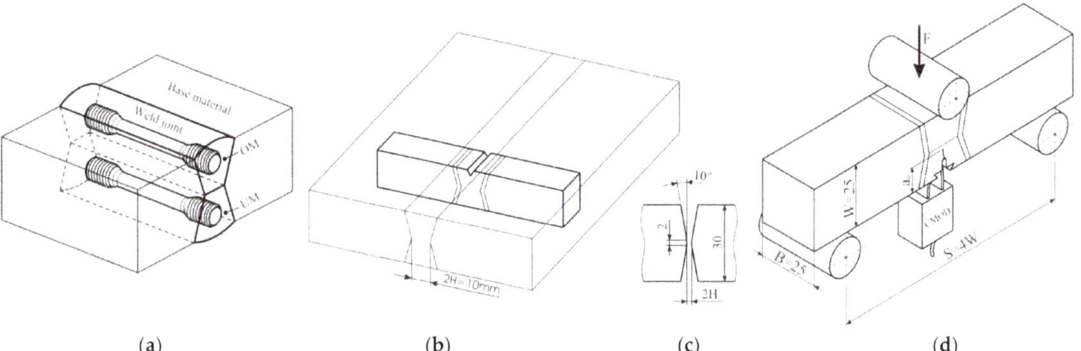

Figure 7. Tensile and fracture toughness specimens: (**a**) position and orientation of tensile specimens; (**b**) fracture toughness specimen notch orientation; (**c**) weld arrangement; (**d**) three point bending specimen SE(B) for fracture toughness testing.

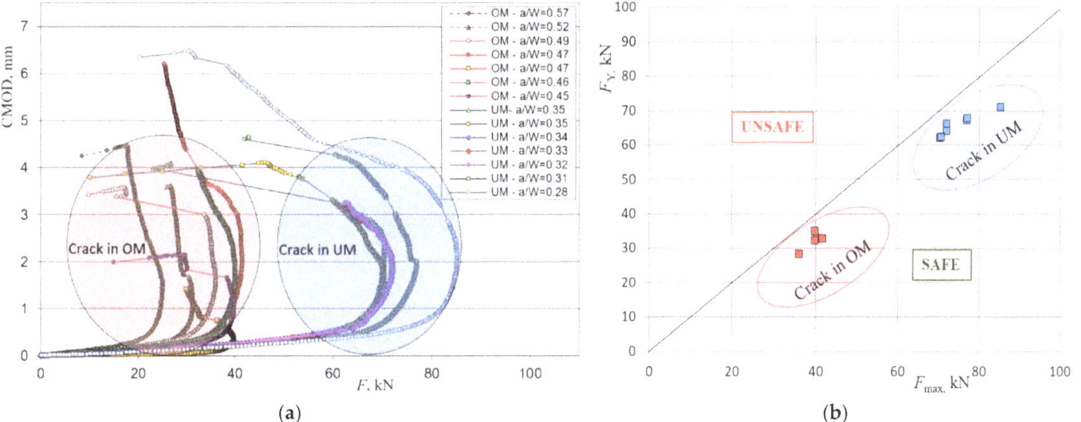

Figure 8. Experimentally obtained results: (**a**) loading curves for specimens with a crack in the OM and UM part of the weld; (**b**) comparison of experimentally obtained maximum load with a yield load obtained by numerical analysis for a crack in the OM and UM part of the weld.

From the loading curve plots, it can be seen that every specimen for each configuration shows a certain period of stable crack propagation and reaches a maximum load, followed by a load decrease and unstable crack propagation. The difference in slopes in the diagrams and values of maximum load were due to different initial crack lengths and position of the crack (OM or UM part of the weld), and a separation of curves for the crack located in the OM part of the weld and advancing to the UM and vice versa can be noted.

The maximum load was determined for each specimen and compared with the yield load solutions obtained by expressions (2) and (3). This comparison is presented in Figure 8b with a yield load versus maximum load plot. Grouping of results is noted for all specimens with a crack in the OM part of the weld as well as with a crack in the UM part of the weld. This was due to relatively similar crack lengths and location of the crack either in the OM or in the UM part of the weld. Yield loads generated from numerically obtained mathematical models for the crack located in the OM and UM part of the weld were lower and conservative enough compared to experimental results. However, it is likely that even less conservative results could be obtained if yield load solutions were implemented as input parameters for evaluating the welded component using structural integrity assessment procedures.

6. Conclusions

As a result of a systematic numerical study, the yield load solutions for SE(B) specimens with a heterogeneous I-shaped weld with an equal share of over-matched and under-matched metal in a welded joint were obtained. Comparing the numerically obtained results in terms of crack position, it can be concluded that a heterogeneous welded joint with a crack in the under-matched metal shows a higher loading capacity than a welded joint with a crack in the over-matched part of a weld. This indicates that in the context of yield load, the metal in front of the crack has a greater effect on the fracture resistance than the metal in which the crack is located. This effect is more pronounced for welds with lower values of weld slenderness $(W - a)/H$. After processing of the numerically obtained results using a symbolic regression algorithm, solutions for the yield loads were proposed with two models: for the crack in the under-matched and for the crack in the over-matched part of the weld. The models were validated with experimental results and they provided sufficiently conservative results compared to the experiment.

Author Contributions: Conceptualization, P.K. and M.K.; methodology, P.K. and N.G.; validation, P.K., D.K. and N.G.; formal analysis, P.K. and M.K.; investigation, P.K. and M.K.; data curation, P.K.; writing—original draft preparation, P.K.; writing—review and editing, P.K., M.K., D.K. and N.G.; visualization, P.K. All authors have read and agreed to the published version of the manuscript.

Funding: Research is funded by EU project Center of Competences for Advanced Engineering Nova Gradiška (CEKOM NI NG) KK.01.2.203.0011 financed by European Regional Development Fund with support of Ministry of the Economy, Entrepreneurship and Crafts of the Republic of Croatia. Experiments were funded by the Marie Curie Fellow Programme No. MCFI-1999-00314.

Institutional Review Board Statement: Not applicable.

Informed Consent Statement: Not applicable.

Data Availability Statement: Not applicable.

Acknowledgments: Nenad Gubeljak acknowledges the European Commission in the frame of the Marie Curie Fellow Programme with the project titled "An estimation of the integrity of a welded structure regarding Constraint" No. MCFI-1999-00314 for funding at the GKSS Research Centre Geesthacht (today named Helmholtz Zentrum Hereon), Institute of Material and Process Design Germany.

Conflicts of Interest: The authors declare no conflict of interest.

References

1. Chen, Z.; Wang, P.; Wang, H.; Xiong, Z. Thermo-Mechanical Analysis of the Repair Welding Residual Stress of AISI 316L Pipeline for ECA. *Int. J. Press. Vessels Pip.* **2021**, *in press*. [CrossRef]
2. Moskvichev, E. Fracture Assessment of Cracked Welded Structures Considering the Heterogeneity of Welded Joints. *Procedia Mater. Sci.* **2014**, *3*, 556–561. [CrossRef]
3. *Structural Integrity Assessment Procedures for European Industry (SINTAP)*; Project BE95–1426, Final Procedure; British Steel Report; British Steel Ltd.: Rotherham, UK, 1999.
4. Su, L.; Xu, J.; Song, W.; Chu, L.; Gao, H.; Li, P.; Berto, F. Numerical Investigation of Strength Mismatch Effect on Ductile Crack Growth Resistance in Welding Pipe. *Appl. Sci.* **2020**, *10*, 1374. [CrossRef]
5. Milosevic, N.; Sedmak, A.; Jovicic, R. Analysis of Strain Distribution in Overmatching V Groove Weld Using Digital Image Correlation. *Procedia Struct. Integr.* **2018**, *13*, 1600–1604. [CrossRef]
6. Zhu, L.; Tao, X.Y. The Study of Weld Strength Mismatch Effect on Limit Loads of Part Surface and Embedded Flaws in Plate. *Int. J. Press. Vessels Pip.* **2016**, *139–140*, 61–68. [CrossRef]
7. Jeremić, L.; Sedmak, A.; Petrovski, B.; Đordević, B.; Sedmak, S. Structural Integrity Assessment of Welded Pipeline Designed with Reduced Safety. *Teh. Vjesn. Tech. Gaz.* **2020**, *27*, 1461–1466.
8. Zhao, H.-S.; Lie, S.-T.; Zhang, Y. Fracture Assessment of Mismatched Girth Welds in Oval-Shaped Clad Pipes Subjected to Bending Moment. *Int. J. Press. Vessels Pip.* **2018**, *160*, 1–13. [CrossRef]
9. Fan, K.; Wang, G.Z.; Tu, S.T.; Xuan, F.Z. Geometry and Material Constraint Effects on Fracture Resistance Behavior of Bi-Material Interfaces. *Int. J. Fract.* **2016**, *201*, 143–155. [CrossRef]
10. Hertelé, S.; O'Dowd, N.; Minnebruggen, K.V.; Verstraete, M.; De Waele, W. Fracture Mechanics Analysis of Heterogeneous Welds: Validation of a Weld Homogenisation Approach. *Procedia Mater. Sci.* **2014**, *3*, 1322–1329. [CrossRef]

11. Zhao, K.; Wang, S.; Xue, H.; Wang, Z. Effect of Material Heterogeneity on Environmentally Assisted Cracking Growth Rate of Alloy 600 for Safe-End Welded Joints. *Materials* **2021**, *14*, 6186. [CrossRef] [PubMed]
12. Starčevič, L.; Gubeljak, N.; Predan, J. The Numerical Modelling Approach with a Random Distribution of Mechanical Properties for a Mismatched Weld. *Materials* **2021**, *14*, 5896. [CrossRef]
13. Naib, S.; Waele, W.D.; Štefane, P.; Gubeljak, N.; Hertelé, S. Analytical Limit Load Predictions in Heterogeneous Welded Single Edge Notched Tension Specimens. *Procedia Struct. Integr.* **2018**, *13*, 1725–1730. [CrossRef]
14. Yu, Y.; Hu, B.; Gao, M.; Xie, Z.; Rong, X.; Han, G.; Guo, H.; Shang, C. Determining Role of Heterogeneous Microstructure in Lowering Yield Ratio and Enhancing Impact Toughness in High-Strength Low-Alloy Steel. *Int. J. Miner. Metall. Mater.* **2021**, *28*, 816–825. [CrossRef]
15. Xue, H.; Wang, Z.; Wang, S.; He, J.; Yang, H. Characterization of Mechanical Heterogeneity in Dissimilar Metal Welded Joints. *Materials* **2021**, *14*, 4145. [CrossRef] [PubMed]
16. Predan, J.; Gubeljak, N.; Kolednik, O. On the Local Variation of the Crack Driving Force in a Double Mismatched Weld. *Eng. Fract. Mech.* **2007**, *74*, 1739–1757. [CrossRef]
17. Kim, Y.-J.; Schwalbe, K.-H. Compendium of Yield Load Solutions for Strength Mis-Matched DE(T), SE(B) and C(T) Specimens. *Eng. Fract. Mech.* **2001**, *68*, 1137–1151. [CrossRef]
18. Kim, Y.-J.; Schwalbe, K.-H. Mismatch Effect on Plastic Yield Loads in Idealised Weldments I. Weld Centre Cracks. *Eng. Fract. Mech.* **2001**, *68*, 163–182. [CrossRef]
19. Kim, Y.-J.; Schwalbe, K.-H. Mismatch Effect on Plastic Yield Loads in Idealised Weldments II. Heat Affected Zone Cracks. *Eng. Fract. Mech.* **2001**, *68*, 183–199. [CrossRef]
20. Kim, S.-H.; Han, J.-J.; Kim, Y.-J. Limit Load Solutions of V-Groove Welded Pipes with a Circumferential Crack at the Centre of Weld. *Procedia Mater. Sci.* **2014**, *3*, 706–713. [CrossRef]
21. Kozak, D.; Gubeljak, N.; Konjatić, P.; Sertić, J. Yield Load Solutions of Heterogeneous Welded Joints. *Int. J. Press. Vessels Pip.* **2009**, *86*, 807–812. [CrossRef]
22. Konjatić, P.; Kozak, D.; Gubeljak, N. The Influence of the Weld Width on Fracture Behaviour of the Heterogeneous Welded Joint. *Key Eng. Mater.* **2011**, *488*, 367–370. [CrossRef]
23. Hao, S.; Schwalbe, K.-H.; Cornec, A. The Effect of Yield Strength Mis-Match on the Fracture Analysis of Welded Joints: Slip-Line Field Solutions for Pure Bending. *Int. J. Solids Struct.* **2000**, *37*, 5385–5411. [CrossRef]
24. Ford, H.; Alexander, J.M. *Advanced Mechanics of Materials*; Longman: London, UK, 1963.
25. Manjgo, M.; Medjo, B.; Milović, L.; Rakin, M.; Burzić, Z.; Sedmak, A. Analysis of Welded Tensile Plates with a Surface Notch in the Weld Metal and Heat Affected Zone. *Eng. Fract. Mech.* **2010**, *77*, 2958–2970. [CrossRef]
26. Zhang, Z.L.; Thaulow, C.; Hauge, M. Effects of Crack Size and Weld Metal Mismatch on the HAZ Cleavage Toughness of Wide Plates. *Eng. Fract. Mech.* **1997**, *57*, 653–664. [CrossRef]
27. ANSYS. Ansys | Engineering Simulation Solutions. 2021. Available online: https://www.ansys.com/ (accessed on 16 November 2021).
28. Jin, Y.; Wang, T.; Krokhin, A.; Choi, T.Y.; Mishra, R. Ultrasonic elastography for nondestructive evaluation of dissimilar material joints. *J. Mater. Process. Technol.* **2022**, *299*, 117301. [CrossRef]
29. Wang, T.; Mishra, R. Effect of Stress Concentration on Strength and Fracture Behavior of Dissimilar Metal Joints. In *Friction Stir Welding and Processing X*; Springer: Cham, Switzerland, 2019.
30. TuringBot Software. Sao Paulo, Brazil. 2021. Available online: https://turingbotsoftware.com/ (accessed on 16 November 2021).
31. *BS 7448-4*; Fracture Mechanics Toughness Tests—Part 4: Fracture Mechanics Toughness Tests. Method for Determination of Fracture Resistance Curves and Initiation Values for Stable Crack Extension in Metallic Materials. British Standard Institution: London, UK, 1991.

Article

Determination of the Actual Stress–Strain Diagram for Undermatching Welded Joint Using DIC and FEM

Nenad Zoran Milošević [1,*], Aleksandar Stojan Sedmak [2], Gordana Miodrag Bakić [2], Vukić Lazić [3], Miloš Milošević [1], Goran Mladenović [2] and Aleksandar Maslarević [1]

[1] Innovation Center of Faculty of Mechanical Engineering, Belgrade University, 11000 Belgrade, Serbia; mmilosevic@mas.bg.ac.rs (M.M.); amaslarevic@mas.bg.ac.rs (A.M.)

[2] Faculty of Mechanical Engineering, Belgrade University, 11000 Belgrade, Serbia; asedmak@mas.bg.ac.rs (A.S.S.); gbakic@mas.bg.ac.rs (G.M.B.); gmladenovic@mas.bg.ac.rs (G.M.)

[3] Faculty of Engineering, University of Kragujevac, 34000 Kragujevac, Serbia; vlazic@kg.ac.rs

* Correspondence: nmilosevic@mas.bg.ac.rs

Abstract: This paper presents new methodology for determining the actual stress–strain diagram based on analytical equations, in combination with numerical and experimental data. The first step was to use the 3D digital image correlation (DIC) to estimate true stress–strain diagram by replacing common analytical expression for contraction with measured values. Next step was to estimate the stress concentration by using a new methodology, based on recently introduced analytical expressions and numerical verification by the finite element method (FEM), to obtain actual stress–strain diagrams, as named in this paper. The essence of new methodology is to introduce stress concentration factor into the procedure of actual stress evaluation. New methodology is then applied to determine actual stress–strain diagrams for two undermatched welded joints with different rectangular cross-section and groove shapes, made of martensitic steels X10 CrMoVNb 9-1 and Armox 500T. Results indicated that new methodology is a general one, since it is not dependent on welded joint material and geometry.

Keywords: actual stress–strain diagram; undermatching weld; martensitic steel; DIC; FEM

Citation: Milošević, N.Z.; Sedmak, A.S.; Bakić, G.M.; Lazić, V.; Milošević, M.; Mladenović, G.; Maslarević, A. Determination of the Actual Stress–Strain Diagram for Undermatching Welded Joint Using DIC and FEM. *Materials* **2021**, *14*, 4691. https://doi.org/10.3390/ma14164691

Academic Editor: Sergei Yu Tarasov

Received: 22 June 2021
Accepted: 12 August 2021
Published: 20 August 2021

Publisher's Note: MDPI stays neutral with regard to jurisdictional claims in published maps and institutional affiliations.

1. Introduction

The tensile diagram, commonly used in practice, is called engineering stress–strain diagram, with both stress and strain defined with respect to the initial, cross-section A_0 and gauge length l_0. For many engineering problems this approximation is good enough, because stresses and strains are close to their true values, as long as contraction and plastic strains are not significant. Anyhow, in the opposite case, true stress–strain diagram is a better option. In its simplest form, true stress and strains are defined as follows, [1]:

$$\sigma_{\mathrm{t}} = \frac{F}{A} = \sigma_{\mathrm{eng}}\left(1 + \varepsilon_{\mathrm{eng}}\right) \tag{1}$$

$$\varepsilon_{\mathrm{t}} = \frac{\Delta l}{l} = ln\left(1 + \varepsilon_{\mathrm{eng}}\right) \tag{2}$$

where σ_{t} and ε_{t} denote the so-called true stress and strain, respectively, F is the acting normal force, A current cross-section, which takes into account the contraction, Δl elongation, l current referent length, $l = l_0 + \Delta l$, l_0 initial length, while σ_{eng} and $\varepsilon_{\mathrm{eng}}$ denote engineering stress and strain, respectively. It should be noted that terms true stress and strain are used here to emphasize the difference with respect to engineering stress and strain, and should not be understood literally. As a matter of fact, modifications of these equations have been in the focus of many researchers for the past few decades.

To start with, based on the fact that contraction is not the only contribution to true stress, couple of other formulas have been proposed, like the formula for equivalent true stress, as defined by Bridgman, [2]:

$$\sigma_{eq} = \frac{\sigma_t}{C_B} \tag{3}$$

where C_B is the correction factor:

$$C_B = \left[\left(1 + \frac{2R}{a}\right)^{1/2} ln\left\{ 1 + \frac{a}{R} + \left(\frac{2a}{R}\right)^{1/2} \left(1 + \frac{a}{2R}\right)^{1/2} \right\} - 1 \right] \tag{4}$$

with a and R representing the ligament and the radius of curvature at the site of contraction.

The same approach is used by Ostsemin [3], with a different correction factor C_O:

$$\sigma_{eq} = \frac{\sigma_t}{C_O} \tag{5}$$

$$C_O = \left(1 + \frac{a}{5R}\right) \tag{6}$$

In [3], a procedure is suggested for calculating the correction for neck formation for round and plane specimens made of homogeneous material. Other correction factors were used in [4] for deriving equivalent stress–strain curve with axisymmetric notched tensile specimens, with experimental verification and good agreement with the Bridgman correction at large strains. Another approach is based on equivalent strain, as defined by Scheider [5]:

$$\bar{\varepsilon} = \sqrt{\frac{4}{3}\left(\varepsilon_x^2 + \varepsilon_x \varepsilon_y + \varepsilon_y^2\right)} \tag{7}$$

leading to:

$$\sigma = \frac{F}{A_0}e^{(\varepsilon_x)} \tag{8}$$

By measuring the mean value of axial strain, formula for the true stress was obtained, [5]:

$$\sigma_t = \frac{F}{A_0}e^{(\bar{\varepsilon}_x)} \tag{9}$$

One should notice that homogeneous material with rectangular cross-section was analyzed in [4,6], where the tensile properties of FH550 and X80 steels were investigated using rectangular cross-section specimens with different thicknesses, respectively.

Tensile diagrams for welded joints have been determined in [7], using novel methods for determining true stress–strain curves for homogenous materials with rectangular cross-section and weldments with round cross-section. In the first case, the relation between the total area reduction and the thickness reduction was derived, consisting of three parts—geometry function, material function, and basic necking curve. In the latter case the central idea was to force plastic deformation at a notch in the material zone of interest, and to obtain the true stress–strain curve of that material zone from the recorded load versus diameter reduction curve.

The same topic was considered in [8], but for different shape of welded joint, the so-called tailor-welded blank weldment. It was concluded that the predicted strain distributions were in good agreement with the measured ones, thus demonstrating the validity of the proposed experimental method to accurately determine the true stress–strain values of the weldment.

More conventional, notched cross weld tensile testing for determining true stress–strain curves for weldments was considered in [9], whereas a method for determining material's equivalent stress–strain curve with any axisymmetric notched tensile specimens without Bridgman correction was considered in [10]. Further in [11] the stress–strain relation for the weld metal is determined through experimental investigations of round tensile specimens. The true stress–strain curve was developed by using the modified version of the weighted average method. Yet another overmatched welded joint was considered in [12], where mechanical behavior with planar type laminations in the base metal (BM), heat-affected zone (HAZ), and welding bead (WB) was studied. By using HV data, an equivalent true stress–strain curve in the HAZ was estimated, based on corresponding hardness value obtained from the BM and WB. In [13] a method to determine the mechanical properties for the weldment of two dual phase (DP) steels is discussed. Inverse numerical simulation was used to simulate the indentation tests to determine and verify the parameters of a nonlinear isotropic material model for the weldment. Results are presented for tensile tests on smooth, notched, and notched-welded specimens. It was shown that the yield and tensile strengths of the notched specimens are higher than the strength of the smooth specimens of the base material due to the additional notch stresses. Similar research is presented in [14], where the microstructure, macro and micro-mechanical properties of dissimilar A302/Cr5Mo were investigated by metallographic experiments, tensile and nanoindentation tests. Based on inversion analysis, elastoplastic properties were estimated for parent metal, weld metal, as well as fine and coarse grain heat-affected zones.

In neither case, presented here, material heterogeneity of a welded joint was not taken into account if a weldment cross-section was rectangular at the same time. The only such a case known to these authors is the welded joint with true stress–strain curves obtained in a special iterative procedure for all local zones (base metal—BM, weld metal—WM, heat-affected zone—HAZ), have different properties, as shown in a series of papers, [15–17]. Anyhow, the iterative procedure presented in [15–17] is not an option here, since it does not lead directly to the result and requires both numerical analysis and experimental testing, not only to verify numerical results, but also to obtain them.

Here, attention is focused to the so-called undermatched welded joint, meaning that the yield stress is lower in a weld metal than in a base metal. One should notice that the plastic strain in undermatched weld metal will appear even with relatively low level of loading, not only due to lower yield stress, but also due to stress concentration, as shown in [18,19]. Once plastic strain becomes significant, cross-section is changed and contraction becomes important, although not the only factor affecting the stress increase. Namely, as it will be shown in this paper, the stress concentration is equally important for this analysis. Therefore, we will use the term actual for the stress–strain diagram exclusively for the case when the stress concentration is taken into account, in addition to contraction.

Toward this aim, one important issue tackled here is the true stress evaluation, which is based on Equation (1), and on contraction values measured by using DIC. As it is shown in this paper, there are significant differences between analytical and measured values of contraction, leading to different true stress–strain curves. For that reason, the term true stress–strain curve is used here for curves obtained by using DIC, whereas the curves obtained by using Equation (1) only are referred to as "true" stress–strain curves. Taking this difference into account, the actual stress–strain curves, as presented here, are based on true stress–strain curves obtained by using DIC, and finally, corrected for the stress concentration.

One should notice that this procedure is a general one, since it will be shown that it does not dependent on welded joint materials and geometry, so it can be applied to overmatched welded joints, as well. Anyhow, since the contraction and plastic strain in that case will be shifted to the base metal, there is almost no practical interest for such an analysis from the point of view of welded joints.

In this work the actual stress–strain diagrams of undermatched welded joints with rectangular cross-section, made of martensitic steel X10 CrMoVNb 9-1 and martensitic armored steel Armox 500T are determined. The goal was to check if different levels of under-dermatching and different shapes of cross-section, as well as different geometry of welded joint, affect actual stress–strain curve, determined by using formulas proposed in this paper. During the experiment, strains were measured in three dimensions using 3D DIC and software Aramis, to evaluate contraction of rectangular cross-section, i.e., to calculate the current cross section of a specimen, so that true stress–strain diagram can be obtained. Finally, correction for the stress concentration is made, using analytical expressions introduced in [20] and verified by comparison with the results of finite element analysis, but only in the case of one material (Armox 500T) and one geometry (specimen P1-1).

Manuscript structure, after the introduction, comprises materials and methods, results, discussion, and conclusions.

2. Materials and Methods

Rectangular test specimens are made of martensitic steel X 10CrMoVNb 9-1 (1.4903–by EN 10216) cut from a pipe, and martensitic steel Armox 500T (SSAB, OxelÖsund, Sweden), cut from a plate. In both cases, a combination of TIG and MMA welding process was used for pipe and plate welding. In both cases S Ni 6082 (EN ISO 18274) was used as filler material for the root and hot pass, and filler material E 19.12.3 Nb R 26 (ISO 3581) was used for filling passes. Chemical compositions of base and filler metals are shown in Tables 1 and 2, respectively.

Table 1. Chemical compositions of base metals.

[%]	C	Mn	Si	Ni	Cr	Mo	B	Cu	V	Ti	Zr	Al_{tot}	Nb	N	P	S
1.4903	0.08–0.12	0.3–0.6	0.2–0.5	≤0.4	8–9.5	0.85–1.05	/	≤0.3	0.18–0.25	/	0.01	≤0.04	0.06–0.1	0.03–0.07	<0.02	<0.01
Armox 500T	0.32	1.2	0.4	1.8	1.0	0.7	0.005	/	/	/	/	/	/	/	0.01	0.003

Table 2. Chemical compositions of the filler metals.

[%]	C	Si	Mn	Cr	Ni	Mo	Nb	Cu	Ti	P	S
S Ni 6082	max 0.01	max 0.1	3.2	20.8	72.9	/	2.5	max 0.1	0.3	0.003	0.001
E 19.12.3 Nb R 26	0.02	0.9	0.7	18.0	12.0	2.7	0.4	max 0.5	/	0.02	0.02

From Table 1 it can be concluded that the base metals used, although both of martensitic microstructure, have significantly different chemical compositions. This is the case because the martensitic microstructure is not obtained in the same way. For 1.4903 steel, martensite was achieved by alloying and consequent heat treatment, whereas for Armox 500T increased carbon content was used, as well as the heat treatment. Materials will not behave in the same way under loading, and this can be concluded by comparing the mechanical properties presented in Table 3 for the base metals and in Table 4 for the filler metals. Materials with different mechanical properties are used to find out if undermatching level affects the proposed formula for stress evaluation. Namely, as one can see from Tables 3 and 4, the undermatching coefficient, defined the ratio between weld metal and base metal yield stress ($R_{p0,2}$), is significantly different, circa 0.9 for steel 1.4903 (400/450) and circa 0.32 (400/1250) for Armox 500T.

Table 3. Mechanical characteristics of the base metals (BM).

BM	Yield Stress [MPa] min	Tensile Strength [MPa]	A [%] min
1.4903	450	630–830	19
Armox 500T	1250	1450–1750	8

Table 4. Mechanical characteristics of the filler metals (FM).

FM	Yield Stress [MPa]	Tensile Strength [MPa]	A_5 [%]	KV [J], 20 °C
S Ni 6082	min 400	min 620	min 35	min 150
E 19.12.3 Nb R 26	min 400	min 590	min 30	min 47

Test specimens were made with "V" joint for 1.4093 steel and with "X" joint for Armox 500T, as shown in Figure 1. Dimension ratios for C1 specimens (steel 1.4903) are 8/10 = 0.8, and for P1 specimens (Armox 500T) are 7.4/7.5 = 0.99, which is practically square. Different shapes of the specimen cross-sections and grooves are also used to find out eventual effects of welded joint geometry on the proposed formulas for stress evaluation.

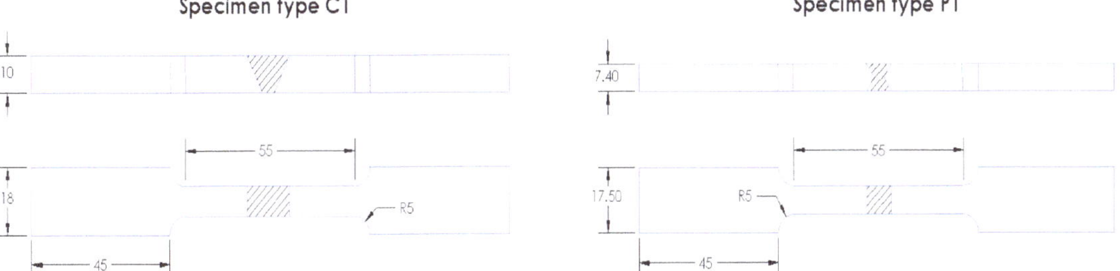

Figure 1. Dimensions [mm] of the specimens for steel 1.4903 (C1) and for Armox 500T (P1).

Digital image correlation (DIC) is a powerful non-contact technique for measuring surface displacement/strain fields, [21]. Simple geometric shapes can be treated by 2D analysis, while more advanced, 3D analysis, should be used for more complex geometric shapes, including welded joints, as applied and presented in [22–24]. The force during the experiment was controlled by strain, with the rate 2 mm/min. Setup of the experiment with the position of cameras is shown in Figure 2. Using DIC method with two cameras (3D deformation measurement) and the Aramis software (Version 2M, GOM GmbH, Braunschweig, Germany) the current cross-section area can be determined. Accuracy of this method for strain measurement is very high, in order of micrometers, so it is a suitable method for the experiment performed here.

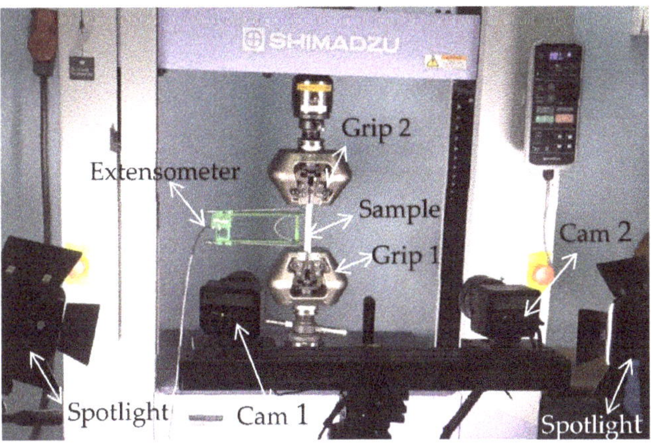

Figure 2. Setting up an experiment with the position of the cameras and with the extensometer set.

Finite element method (FEM) is nowadays a widely accepted numerical tool to get stress and strain distribution for many engineering problems, including elastic-plastic analysis of welded joints, even in the presence of cracks, and for other complex problems, [25,26].

Here, 3D FEM is used to evaluate stress concentration. Mesh was made with 3D linear elements, C3D8, with 8 nodes, with decreasing size in the weld metal down to 0.4×0.2 mm, as shown in Figure 3, where one example of meshes deformed in weld metal is given. One quarter of specimen was modeled due to two planes of symmetry and appropriate boundary conditions applied (one rotation and two translations fixed). Load is defined as the negative pressure, according to the force applied and remote cross-section. More detailed description is given in [20].

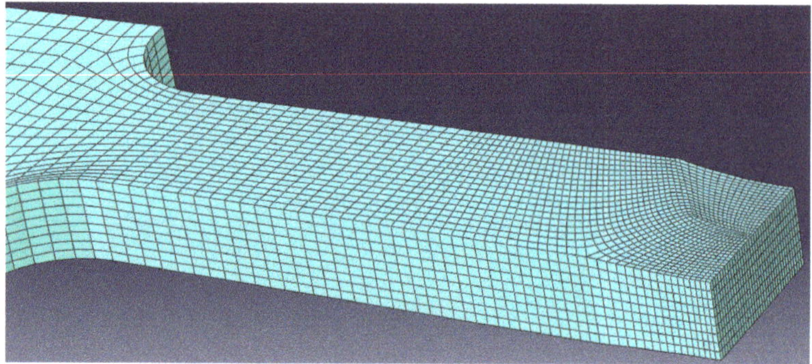

Figure 3. FE model of specimen P1-1 with deformed weld metal.

3. Results

Typical result for strain measurement by DIC is shown in Figure 4, as obtained by the post-processing, using software Aramis.

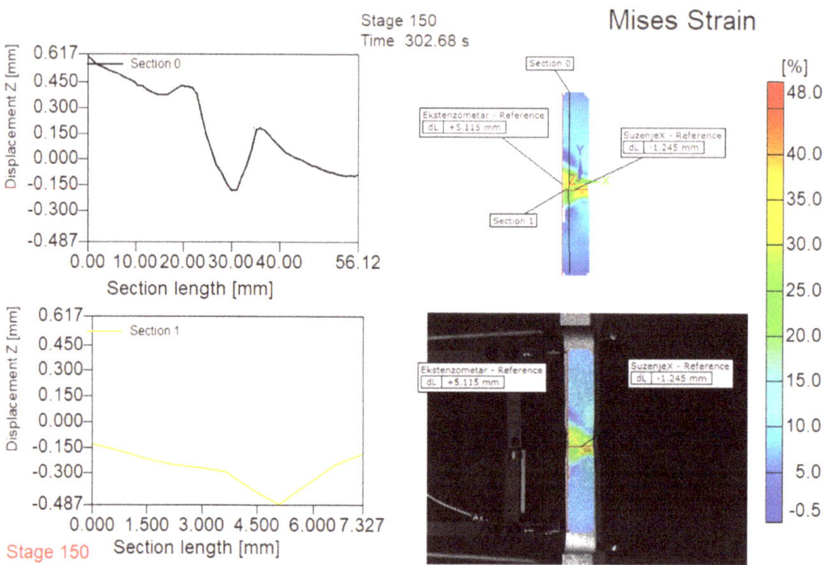

Figure 4. Analysis of changes in the characteristic dimensions of the specimen C1-1.

The current cross-section area of the specimen was calculated using data obtained by Aramis, as shown in Figure 5 for specimen P1-1. One of the sides was actually measured, the opposite one taken as the mirror image, and two remaining are obtained by rotating the measured one for 90° and −90°.

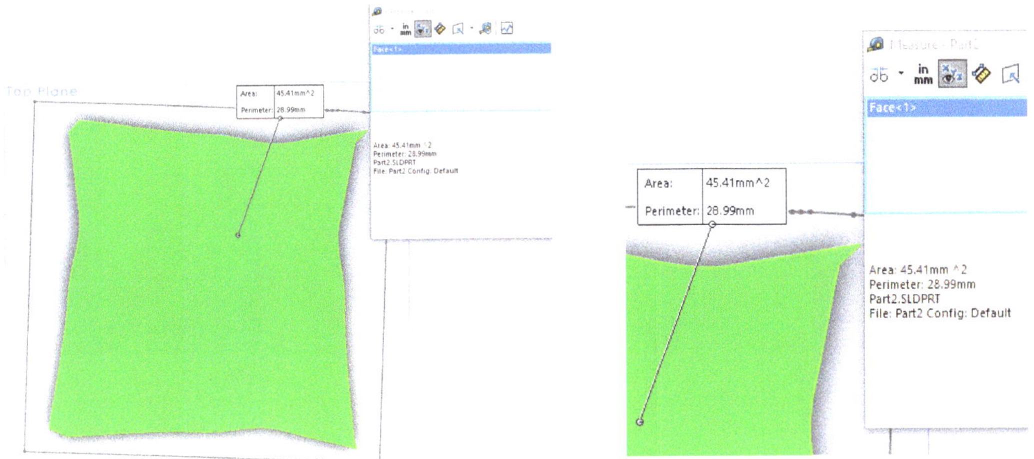

Figure 5. Initial (marked by line) and final (green) cross-section area of specimen P1-1.

Figure 6 shows three stress–strain diagrams for both specimens, types C1 and P1, including engineering diagram, obtained by standard tensile test, marked in black. Remaining two diagrams represent true stress–strain curves, one determined according to Equations (1) and (2), marked in red, and the other one determined using measured cross-section areas of the specimen by DIC, marked in blue. One can see that the true stress is increased, if contraction measured by DIC (Figure 5) is taken as relevant. This is why red curves in Figure 6 are marked as "true" and blue ones as true.

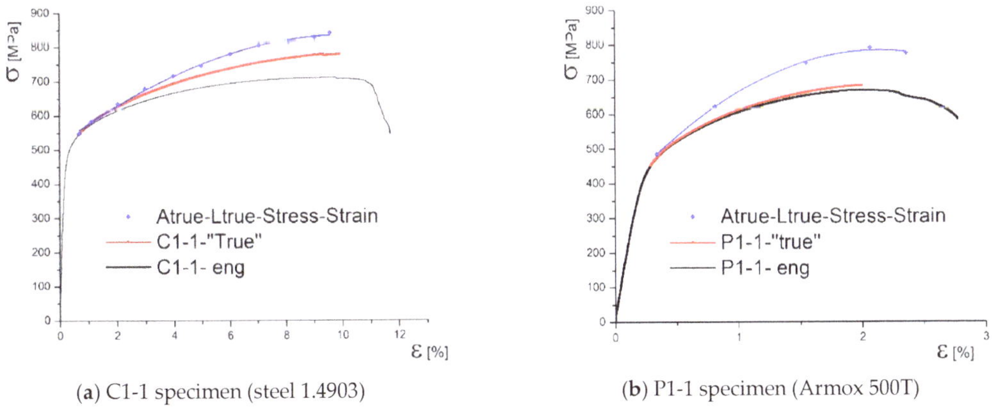

(**a**) C1-1 specimen (steel 1.4903)   (**b**) P1-1 specimen (Armox 500T)

Figure 6. Comparison of true and engineering diagrams calculated by DIC method.

Results of FEM calculation are shown in Figure 7 for specimen C1-1 as an example of the procedure applied. Results for C1-1 specimen, with deformed weld metal according to strains and contraction obtained by DIC, show equivalent stress distribution, Figure 7a, and normal stresses distribution, Figure 7b, for the applied load 4 KN, producing remote tensile stress 100 MPa in the narrow part of the specimen, away from the welded joint area.

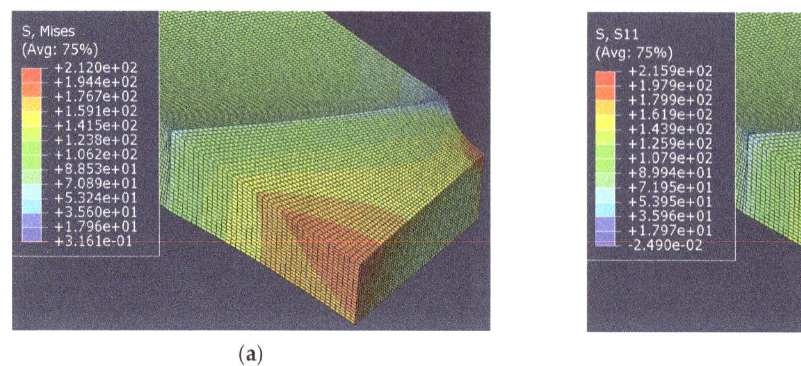
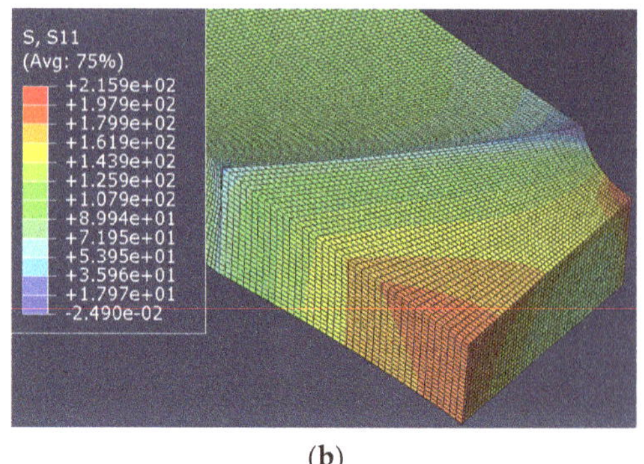

(a)   (b)

Figure 7. Stress distribution in specimen C1-1 with deformed weld metal: (**a**) equivalent stresses, (**b**) normal stresses, given in MPa.

From Figure 7 it can be concluded that the difference between maximum Misses equivalent stress and maximum normal stress is just 3.91 MPa (215.9–212 MPa) or 1.84%. This leads to the conclusion that the equivalent stress is not the dominant parameter for stress increase, but it is rather the stress concentration due to contraction. To calculate the actual stress with the stress concentration taken into account, the authors propose the following equations:

$$\sigma_{\max}^{\text{actual}} = \sigma_{\text{T}} C_{\text{NM}} \tag{10}$$

where C_{NM} is the stress concentration factor and σ_{T} is calculated as:

$$\sigma_{\text{T}} = \frac{F}{A_{\text{current}}} \tag{11}$$

Stress concentration factor C_{NM} can be separated into two factors, as follows:

$$C_{\text{NM}} = C_{\text{ZS}} + C_{\text{EP}} \tag{12}$$

where C_{ZS} takes into account the welded joint geometry and C_{EP} stands for reduction of thickness. According to [22], C_{ZS} can be expressed for point 1, as follows:

$$C_{\text{ZS}_1} = 1 + \frac{b_1}{2(R_1 + b_1)} \tag{13}$$

where b_1 and R_1 are defined in Figure 8 for two characteristic points in a weld metal, together with their counterparts, b_2 and R_2, used for calculating C_{ZS} for point 2.

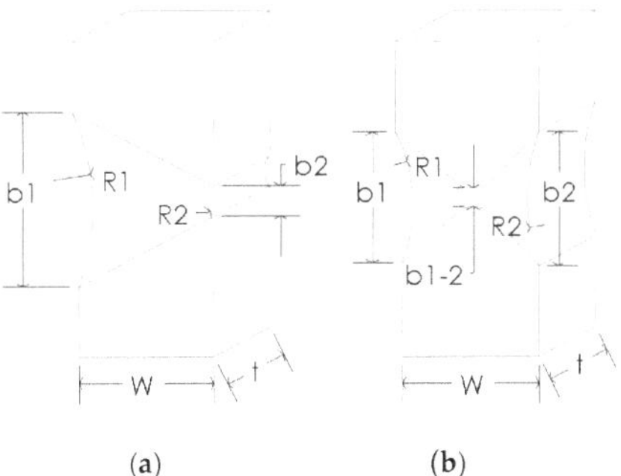

Figure 8. Characteristic dimensions of weld metals: (**a**) V shape, (**b**) X shape.

Likewise, C_{EP} can be defined as, [22]:

$$C_{EP} = \frac{\Delta t}{2W_0} = \frac{\Delta t / t_0}{2W_0 / t_0} \tag{14}$$

where t_0 and W_0 are initial values of thickness t and width W, Figure 8. Therefore, the final expression for the stress concentration factor is:

$$C_{NM} = \left(1 + \frac{b}{2(R+b)} + \frac{\Delta t}{2W_0}\right) \tag{15}$$

The current cross-sectional area of the specimen ($A_{current}$) was calculated using the data obtained by the DIC.

In the further analysis, numerical verification of coefficients for specimens C1-1 and P1-1 is shown for strains immediately before the fracture:

- Specimen C1-1

C1-1	$t_0 = 8$ [mm]	$W_0 = 10$ [mm]	$F = 48342.18$ [N]	$A_{current} = 63.23293$ [mm^2]	$\sigma_T =$ 764.50957 [MPa]	$\Delta t =$ 1.0487 [mm]
Point 1	$b_1 = 18.866$ [mm]	$R_1 = 58.0536$ [mm]	$C_{NM_1} = 1.17507091$	$\sigma_{max_1}^{actual} = \sigma_T C_{NM_1} = 898.353$ [MPa]		
Point 2	$b_2 = 8.2363$ [mm]	$R_2 = 11.3768$ [mm]	$C_{NM_2} = 1.262405968$	$\sigma_{max_2}^{actual} = \sigma_T C_{NM_2} = 965.121$ [MPa]		

- Specimen P1-1

P1-1	$t_0 = 7.5$ [mm]	$W_0 = 7.4$ [mm]	$F = 35885.96$ [N]	$A_{current} = 46.20984$ [mm^2]	$\sigma_T =$ 776.587 [MPa]	$\Delta t =$ 0.412963 [mm]
Point 1	$b_1 = 10.2741$ [mm]	$R_1 = 35.14171$ [mm]	$C_{NM_1} = 1.168917692$	$\sigma_{max_1}^{actual} = \sigma_T C_{NM_1} = 907.766$ [MPa]		
Point 2	$b_2 = 10.2595$ [mm]	$R_2 = 35.042$ [mm]	$C_{NM_2} = 1.169041427$	$\sigma_{max_2}^{actual} = \sigma_T C_{NM_2} = 907.862$ [MPa]		

The values obtained in ABAQUS for the quarter of the specimen C1-1 and P1-1 at the characteristic points (1 and 2) are shown in Figure 9. Stress for specimen C1-1, the maximum equivalent stresses (von Misses) are:

$$S_{Misses_1} = 901.605 \text{ MPa, i.e., } S_{Misses_2} = 1004.67 \text{ MPa} \tag{16}$$

For specimen P1-1, the maximum equivalent stresses (Misses) by Abaqus are:

$$S_{Misses_1} = 884.737 \text{ MPa, i.e., } S_{Misses_2} = 884.888 \text{ MPa} \tag{17}$$

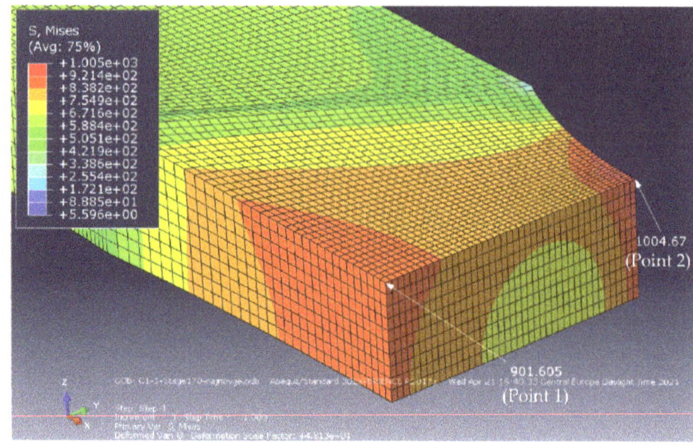

(**a**)

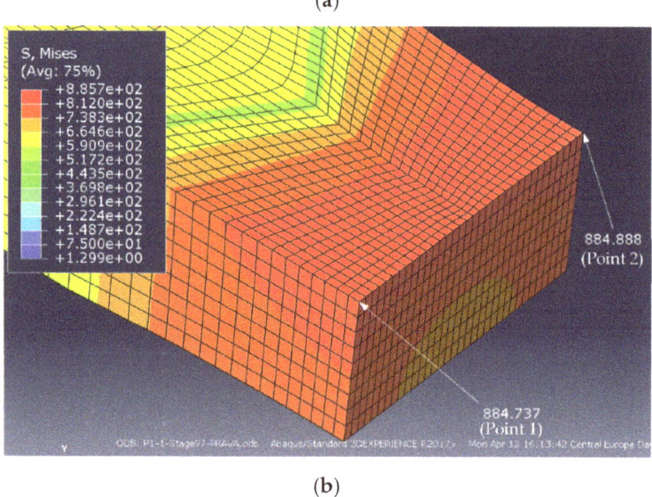

(**b**)

Figure 9. Maximum equivalent stresses in MPa at the characteristic points for specimens: (**a**) C1-1, (**b**) P1-1.

The equivalent stress values, obtained by ABAQUS and the stresses calculated by the formulas (10)–(15), are given in Table 5. One should notice difference between stress values in points 1 and 2 for specimen C1-1 and almost the same stress values in these two points for specimen P1-1.

Table 5. Comparison of the maximal stresses for the specimen C1-1 and P1-1.

Specimen	Calculated $\sigma_{\max_1}^{\text{actual}}$	Abaqus Point 1	Difference [%]	Calculated $\sigma_{\max_2}^{\text{actual}}$	Abaqus Point 2	Difference [%]
C1-1	898.4	901.6	0.36	965.1	1004.7	4.1
P1-1	907.8	884.7	2.6	907.9	884.9	2.6

In Figures 10 and 11, actual, true, and engineering stress–strain diagrams are presented for the specimen C1-1 and for the specimen P1-1, respectively.

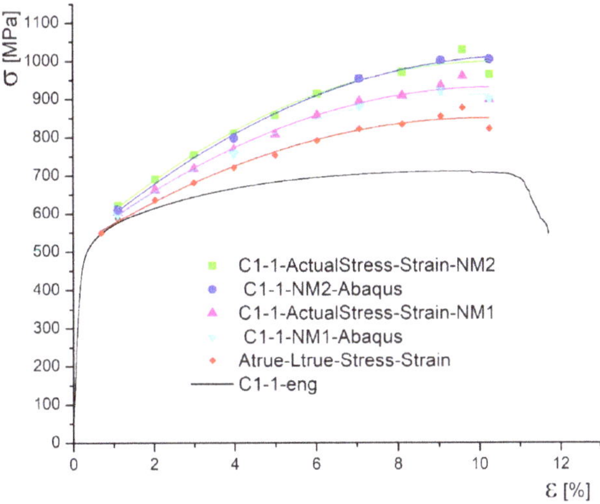

Figure 10. Actual stress–strain diagrams for the specimen C1-1.

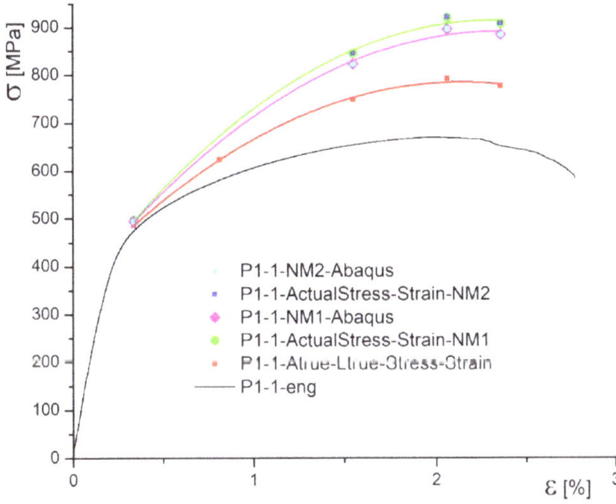

Figure 11. Actual stress–strain diagrams for the specimen P1-1.

4. Discussion

In this research the new methodology for true stress–strain curves are applied to undermatched welded joints made of different base metals, with different geometries (cross section and groove shape). One should notice that both base metals, used in this research, are low plasticity materials, especially Armox 500T (elongation A = 8%). Therefore, using only Equations (1) and (2) for determining the true stress–strain diagram produced questionable result, since the force drop is followed by the stress drop, as shown in Figure 6 for both base metals. Thus, the real contraction, as measured by 3D DIC, should be also taken into account, providing more realistic true stress–strain curves for both base metals, also shown in Figure 6. As already mentioned, at this stage of development, one side of the specimen was actually measured, and the opposite one taken as the mirror image, while the remaining two sides are obtained by rotation. Anyhow, this issue will be tackled in future work by using at least four cameras to measure the two sides, and get the other two as mirror images. Measuring all the sides is probably too complicated, but it will be considered, as well.

Anyhow, in addition to previous, stress concentration due to geometry change should also be taken into account. Toward this end, new analytical expressions, i.e., formulas (10)–(15) have been introduced in the scope of this research, and verified by using the FEM. This was enabled by using results for strains and contraction, as obtained by DIC, to form FE models with different geometries of weld metal for different load levels, as explained in more detail in [20] using one base metal and one welded joint geometry. Here, this methodology is applied to both base metal and welded joint geometries to investigate eventual effects on actual stress–strain curves.

From Figures 10 and 11 one can see that actual stresses $\sigma_{\max_1}^{\text{actual}}$ and $\sigma_{\max_2}^{\text{actual}}$ differ in specimen C1-1, while in the specimen P1-1 they are almost the same. Clearly, this is the effect of joint shape, since V joint (specimen C1-1) has different dimensions b_1 and b_2, and thus different radii of curvature R_1 and R_2, leading to different stress concentration factors, as well. For the specimen P1-1, difference between $\sigma_{\max_1}^{\text{actual}}$ and $\sigma_{\max_2}^{\text{actual}}$ is negligible due to the symmetry of joint shape (X), having approximately same values of b_1 and b_2, and radii of curvature, R_1 and R_2, leading to almost the same stress concentration factors.

It is also important to notice that differences in stresses calculated by the proposed formulas (10)–(15) and equivalent stresses obtained by Abaqus for the moment immediately before the fracture, Figure 9, do not exceed 4.10% (specimen C1-1, Table 5). With this in mind, it can be considered that the proposed formulas evaluate the actual stress correctly for different levels of undermatching and different types of weld groove, as well as different shape ratio of the specimen cross section. Therefore, it was proved here that the proposed methodology is a general one, and can be applied to different materials and welded joint geometries.

One should notice that these effects are important for undermatched welded joints, since only in this case plastic strain and stress concentration develop in the weld metal, contrary to the overmatching welded joint, where they shift to the base metal, i.e., out of the critical zones of welded joint. Anyhow, it is still important to analyze overmatching effect in future research, since it is the most often case in practice.

5. Conclusions

The proposed Equations (10)–(15) proved to be sound basis to determine the actual stress–strain diagrams for undermatching the welded joints made of different base metals with different welded joint geometries. Actual stresses obtained by these formulas are in good agreement with the equivalent stresses obtained by Abaqus using finite element meshes constructed according to the geometry obtained by DIC.

It can be concluded that the actual value of the tensile strength of a welded joint is far above the value obtained by the standard tensile testing, presented by engineering stress–strain curves. This difference is a consequence of cross-section contraction and stress concentration in the most deformed zone, being the weld metal in the case of undermatched welded joint.

Cross-section contraction turned out to be an important factor in the case of low plasticity material, as used in this research, since the usual formulas for "true" stress–strain curves provide questionable behavior with drop of stress after maximum tensile force is reached.

The differences in normal and equivalent stress in rectangular specimens are not significant, leading to the conclusion that the dominant effect in rectangular specimens is not triaxial stress state, but the stress concentration due to contraction.

Further analysis should use more ductile material to analyze their behavior with respect to cross-section contraction and stress concentration, as well as other types of welded joints, such as overmatching joints and different welded joint geometries, to suggest eventual corrections to the proposed formulas.

Author Contributions: Conceptualization, N.Z.M. and A.S.S.; methodology, N.Z.M.; software, N.Z.M. and M.M.; validation, N.Z.M., A.S.S. and G.M.B.; formal analysis, N.Z.M. and A.S.S.; investigation, N.Z.M.; resources, V.L., M.M. and G.M.; data curation, N.Z.M., A.S.S.; writing—original draft preparation, N.Z.M.; writing—review and editing, A.S.S. and N.Z.M.; visualization, N.Z.M., A.S.S. and A.M.; supervision, A.S.S. and G.M.B. All authors have read and agreed to the published version of the manuscript.

Funding: The results presented are part of a research supported by MESTD RS by contract 451-03-9/2021-14/200105.

Institutional Review Board Statement: Not applicable.

Informed Consent Statement: Not applicable.

Data Availability Statement: The data presented in this study are available on request from the corresponding author. The data are not publicly available due to [ongoing research].

Conflicts of Interest: The authors declare no conflict of interest.

References

1. Ling, Y. Uniaxial true stress-strain after necking. *AMP J. Technol.* **1996**, *5*, 37–48.
2. Bridgman, P.W. *Studies in Large Plastic Flow and Fracture*; Harvard University Press: Cambridge, MA, USA, 1964. Available online: https://www.hup.harvard.edu/catalog.php?isbn=9780674731349 (accessed on 14 August 2021).
3. Ostsemin, A.A. Stress in the least cross section of round and plane specimens in tension. *Strength Mater.* **1992**, *24*, 298–301. [CrossRef]
4. Tu, S.; Ren, X.; He, J.; Zhang, Z. Experimental measurement of temperature-dependent equivalent stress-strain curves of a 420 MPa structural steel with axisymmetric notched tensile specimens. *Eng. Fail. Anal.* **2019**, *100*, 312–321. [CrossRef]
5. Scheider, I.; Brocks, W.; Cornec, A. Procedure for the Determination of True Stress-Strain Curves from Tensile Tests with Rectangular Cross-Section Specimens. *J. Eng. Mater. Technol.* **2004**, *126*, 70–76. [CrossRef]
6. Yuan, W.; Zhang, Z.; Su, Y.; Qiao, L.; Chu, W. Influence of specimen thickness with rectangular cross-section on the tensile properties of structural steels. *Mater. Sci. Eng. A* **2012**, *532*, 601–605. [CrossRef]
7. Zhang, Z.L.; Ødegård, J.; Thaulow, C. Novel methods for determining true stress strain curves of weldments and homogenous materials. In Proceedings of the 13th European Conference on Fracture (ECF 13), San Sebastian, Spain, 6–9 September 2000.
8. Cheng, C.; Jie, M.; Chan, L.C.; Chow, C. True stress–strain analysis on weldment of heterogeneous tailor-welded blanks—A novel approach for forming simulation. *Int. J. Mech. Sci.* **2006**, *49*, 217–229. [CrossRef]
9. Zhang, Z.L.; Hauge, M.; Thaulow, C.; Ødegard, J. A notched cross weld tensile testing method for determining true stress–strain curves for weldments. *Eng. Fract. Mech.* **2002**, *69*, 353–366. [CrossRef]
10. Tu, S.; Ren, X.; He, J.; Zhang, Z. A method for determining material's equivalent stress-strain curve with any axisymmetric notched tensile specimens without Bridgman correction. *Int. J. Mech. Sci.* **2018**, *135*, 656–667. [CrossRef]
11. Benjamin, W.; Horst, H. Manuela Sander, Experimental and numerical investigation of fracture in fillet welds by cross joint specimens. *Procedia Struct. Integr.* **2016**, *2*, 2054–2067.
12. Fernández-Cueto, M.J.; Capula-Colindres, S.; Angeles-Herrera, D.; Velazquez, J.C.; Méndez, G.T. Analysis of 3D Planar Laminations in a Welded Section of API 5L X52 Applying the Finite Element Method. *Soldag. Inspeção* **2018**, *23*, 17–31. [CrossRef]
13. Javaheri, E.; Lubritz, J.; Graf, B.; Rethmeier, M. Mechanical Properties Characterization of Welded Automotive Steels. *Metals* **2019**, *10*, 1. [CrossRef]
14. Jiang, Y.; Wu, Q.; Zhao, J.; Gong, J. Characterization of elastoplastic properties of dissimilar weld joint of A302/Cr5Mo using the inversion analysis. *Mater. Res. Express* **2019**, *6*, 116552. [CrossRef]
15. Younise, B.; Sedmak, A.; Milosević, N.; Rakin, M.; Medjo, B. True Stress-strain Curves for HSLA Steel Weldment—Iteration Procedure Based on DIC and FEM. *Procedia Struct. Integr.* **2020**, *28*, 1992–1997. [CrossRef]
16. Younise, B.; Rakin, M.; Gubeljak, N.; Medjo, B.; Sedmak, A. Numerical simulation of constraint effect on fracture initiation in welded specimens using a local damage model. *Struct. Integr. Life* **2011**, *11*, 51–56.
17. Younise, B.; Sedmak, A. Micromechanical study of ductile fracture initiation and propagation on welded tensile specimen with a surface pre-crack in weld metal. *Struct. Integr. Life* **2014**, *14*, 185–191.
18. Sedmak, S.; Sedmak, A. Integrity of penstock of hydroelectric power plant. *Struct. Integr. Life* **2005**, *5*, 59–70.
19. Jeremić, L.; Sedmak, A.; Petrovski, B.; Đorđević, B.; Sedmak, S. Structural Integrity Assessment of Welded Pipeline Designed with Reduced Safety. *Teh. Vjesn. Tech. Gaz.* **2020**, *27*, 1461–1466. [CrossRef]
20. Milosevic, N.; Sedmak, A.; Martic, I.; Prokic-Cvetkovic, R. Novel procedure to determine actual stress-strain curves. *Struct. Integr. Life* **2021**, *21*, 37–40.
21. Sedmak, A.; Milošević, M.; Mitrović, N.; Petrović, A.; Maneski, T. Digital image correlation in experimental mechanical analysis. *Struct. Integr. Life* **2012**, *12*, 39–42.

22. Milosevic, M.; Mitrovic, N.; Jovicic, R.; Sedmak, A.; Maneski, T.; Petrovic, A.; Aburuga, T. Measurement of Local Tensile Properties of Welded Joint Using Digital Image Correlation Method. *Chem. Listy* **2012**, *106*, 485–488.
23. Milosevic, N.; Sedmak, A.; Jovicic, R. Analysis of strain distribution in overmatching V groove weld using digital image correlation. *Procedia Struct. Integr.* **2018**, *13*, 1600–1604. [CrossRef]
24. Milosevic, M.; Sedmak, S.; Tatic, U.; Mitrovic, N.; Hloch, S.; Jovičić, R. Digital image correlation in analysis of stiffness in local zones of welded joints. *Teh. Vjesn. Tech. Gaz.* **2016**, *23*, 19–24. [CrossRef]
25. Sedmak, A. Computational fracture mechanics: An overview from early efforts to recent achievements. *Fatigue Fract. Eng. Mater. Struct.* **2018**, *41*, 2438–2474. [CrossRef]
26. Banks-Sills, L.; Sedmak, A. Linear elastic and elasto-plastic aspects of interface fracture mechanics. *Struct. Integr. Life* **2020**, *20*, 203–210.

materials

Article

The Numerical Modelling Approach with a Random Distribution of Mechanical Properties for a Mismatched Weld

Luka Starčevič, Nenad Gubeljak * and Jožef Predan

Faculty of Mechanical Engineering, University of Maribor, Smetanova 17, SI-2000 Maribor, Slovenia; luka.starcevic@student.um.si (L.S.); jozef.predan@um.si (J.P.)
* Correspondence: nenad.gubeljak@um.si

Abstract: The aim of this work was to include a local variation in material properties to simulate the fracture behaviour in a multi-pass mis-matched X-weld joint. The base material was welded with an over and under-match strength material. The local variation was represented in a finite element model with five material groups in the weld and three layers in the heat-affected zone. The groups were assigned randomly to the elements within a region. A three-point single edge notch bending (SENB) fracture mechanics specimen was analysed for two different configurations where either the initial crack is in the over or under-matched material side to simulate experimentally obtained results. The used modelling approach shows comparable crack propagation and stiffness behaviour, as well as the expected, scatter and instabilities of measured fracture behaviour in inhomogeneous welds.

Keywords: weld metals; welded joints; damage mechanics; finite element analysis; crack growth; ductile fracture

Citation: Starčevič, L.; Gubeljak, N.; Predan, J. The Numerical Modelling Approach with a Random Distribution of Mechanical Properties for a Mismatched Weld. *Materials* **2021**, *14*, 5896. https://doi.org/10.3390/ma14195896

Academic Editor: Alessandro Pirondi

Received: 26 August 2021
Accepted: 26 September 2021
Published: 8 October 2021

Publisher's Note: MDPI stays neutral with regard to jurisdictional claims in published maps and institutional affiliations.

1. Introduction

Many researchers [1–6] who deal with numerical simulations of the strength and fracture behaviour of welds are looking for a suitable numerical universal tool to describe as faithfully as possible the behaviour of a weld with a crack. In particular, they focus on crack propagation through different strength weld materials, as, in the case of confirmation of the correctness of these tools, simulations can be performed for different weld shapes and for different materials and different loading methods [7–9].

Welded joints represent heavily inhomogeneous material regions of structures, which result in a local crack driving force and in a crack path deviation, where a crack propagates through different strength regions. The effect is also reflected globally in the force vs. displacement load curve. Many researchers [10–16] have investigated the influence of the material properties' inhomogeneities in a welded joint on fracture behaviour using experimental and numerical methods. They developed an approach for local crack driving force determination based on the configurational force concept [17]. The local crack driving force is calculated by post-processing followed by a classical finite element analysis as the sum of a far-field crack driving force and additional material inhomogeneity term. Many studies have been published for different material inhomogeneity configurations and spatial variations in material properties. They studied the effect of material inhomogeneities for discrete jumps of material properties at the interfaces, as well as continuous variation in properties in biomaterials. In the numerical simulations, they were focused on the point of crack initiation of the stable crack growth [16], where they obtained a good match between the experimental and numerical results, but, in the case of crack growth, they received significant deviations due to the crack deviation from the initial pre-fatigue crack plane.

Globally, distinct strength inhomogeneous welds are repair multi-pass welds in high loaded structures, where the part of the weld with the defect must be removed by grooving and filling with an under-strength filler material. If hidden defects such as pores or non-melted situ occur during repair welding, a crack is initiated in the low-strength weld material and propagates towards the high-strength part of the weld. Globally, two regions with different material properties affect the local crack driving force magnitude and direction, which influence crack growth rate and deviates the crack path from the initial straight. Some fracture instabilities can be caused by the extremely increased local crack driving force by the material inhomogeneity in the cases where the crack propagates from the over to under-strength material, and vice versa, the crack can be arrested by the diminishing local crack driving force in the case where the crack approaches the interface from the lower strength material.

To ensure the structure integrity of the weld in the presence of a crack, it is important to estimate the residual load capacity through the force displacement load curve.

The purpose of the study is to present the numerical simulation results of the load vs. crack mouth opening displacement (CMOD) curve for the propagating cracks through globally and locally inhomogeneous welds, taking into account the local mechanical properties obtained from the standard and mini tensile specimens (MTSs), as well as with the empirical correlation between the microhardness and strength.

The subject of the numerical simulation is the fracture behaviour of a multi-pass inhomogeneous weld consisting of two different filler materials with an initial crack in the under-strength weld part growing towards the over-strength half and with an initial crack in the over-strength weld part growing towards the under-strength weld half.

It is well-known that weld joints have inhomogeneous mechanical properties. These usually appear in multi pass mismatched welds, where the properties are combined in order to achieve the desired fracture behaviour. Typically, the combination of mechanical properties in a mismatched weld are the following: one half under (UM) and one half over (OM) matched weld material, and on both sides of the heat affected zones (HAZs) and base material (BM). Usually, the weld material should be an OM weld metal in order keep the OM material elastic, while plasticity starts in the BM. Therefore, the higher probability to failure is expected in the BM or HAZ than the OM weld metal. The combination of selected weld materials can, therefore, affect the stiffness response and crack paths of the weld joint significantly. Thus, the structural integrity of the cracked mismatched weld joint depends mainly on the fracture toughness of the cracked zone and loading condition [18–20].

The local mechanical properties at the BM, HAZ and inhomogeneous weld should be considered in order to consider the structural integrity for designing the weld structures. In the past, detailed experimental investigations were carried out for multi pass welds to analyse the local mechanical performance inside the weld and HAZ [21–25] in the SENB specimen. It is known that mechanical properties such as spatial yield stresses vary within the hardener and softer zones (HAZs) and the weld region. Therefore, this local material inhomogeneity should be considered in the finite element (FE) simulation, as in the latest approaches [26–28].

However, a sufficient approach for modelling a multi pass weld does not exist yet, and is required to analyse failure potentials (crack paths) for welded structures. With such a model, critical welds inside large structures (pressure vessels, welded components, etc.) can be analysed and optimised to reach damage tolerant behaviour.

2. Materials and Experiments

In our case, we focused on two materials deposited in multi pass "X" -welded joints, with two crack configurations, either the initial crack in the UM (configuration 1) or OM (configuration 2) weld site, according to Figure 1 and Table 1. NIOMOL 490 was used as a base metal (BM), FILTUB 75 as an OM and VAC 60 as UM materials for the weld.

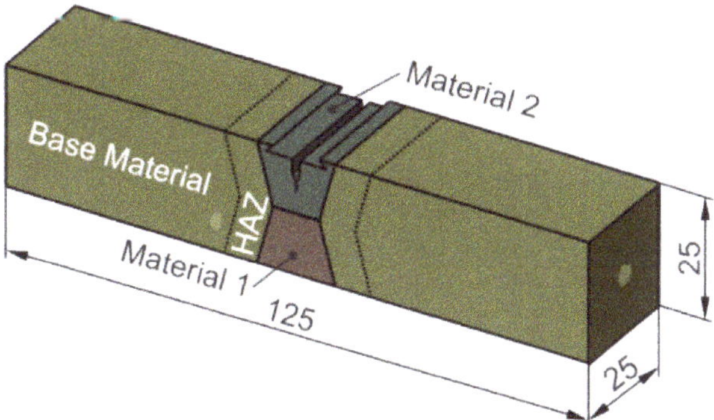

Figure 1. Schematic position of a mechanical notch in a bi-material welded SENB specimen.

Table 1. Tensile mechanical properties.

Configuration	Crack Growth Direction	Material 1	Material 2
I	UM → OM	OM (FILTUB 75)	UM (VAC 60)
II	OM → UM	UM (VAC 60)	OM (FILTUB 75)

The material NIOMOL 490 is a high-strength low-alloy fine grain steel, with retard to coarse grain growing in the heat affected zone. Therefore, NIOMOL 490K has very good weldability and it is possible to welded without preheating with low strength consumable materials, e.g., VAC60. The mechanical properties and chemical compositions of the BM, the OM and UM weld metals are provided in Tables 2 and 3. A flux cord arc welding (FCAW) procedure was applied, and two different tubular wires were selected for welding in order to produce welded joints in over- and under-matched (OM and UM) configurations. The heat input of each weld pass was between 15 and 18 kJ/cm, corresponding to the cooling time between 500 and 800 °C $\Delta t_{8/5}$ = 8–12 s. Such weld metal configurations are common for repairing welding.

Table 2. Tensile mechanical properties.

Material	Lebel	R_{p02} [MPa]	R_m [MPa]	M	Charpy, Kv
Base material	NIOMOL 490	510	650	-	>60 J at −50 °C
Over matched	FILTUB 75	700	780	1.37	>40 J at −50 °C
Under matched	VAC 60	437	556	0.86	>80 J at −50 °C

Table 3. Actual chemical composition (in weight %).

Material	C	Si	Mn	P	S	Cr	Mo	Ni
Base material	0.123	0.33	0.56	0.003	0.002	0.57	0.34	0.13
Over matched	0.040	0.16	0.95	0.011	0.021	0.49	0.42	2.06
Under matched	0.096	0.58	1.24	0.013	0.16	0.07	0.02	0.03

To determine the tensile behaviour, experimental testing of five specimens was carried out to collect the yield strengths R_{p02} and R_m as a reference for the material model development.

The base material properties were kept constant, while the weld metal properties varied. This variation is described by the mismatch factor:

$$M = \frac{\sigma_{YW}}{\sigma_{YB}} \tag{1}$$

where σ_{YW} and σ_{YB} present the yield strength of the weld metal and the yield strength of the base metal, respectively. The weld metal is commonly produced with a yield strength greater than that of the base plate; this case is designated as overmatching (OM) with the mismatch factor $M > 1$. However, an increasing use of high-strength steels forces the fabricator to select a consumable with lower strength to comply with the toughness requirements, which are designated as under-matching (UM), where $M < 1$.

Later, the local variation in the tensile behaviour was tested and analysed in the weld for the UM and OM weld material, as well as the HAZ, by using a set of mini tensile specimens (MTSs). The orientation and position of both set of specimens is shown in Figure 2. MTSs are fabricated by wire spark eroding techniques. MTS testing was performed by uniaxial testing under a constant stroke velocity of 0.1 mm/min and by laser extension measurement, with an initial length of 8 mm.

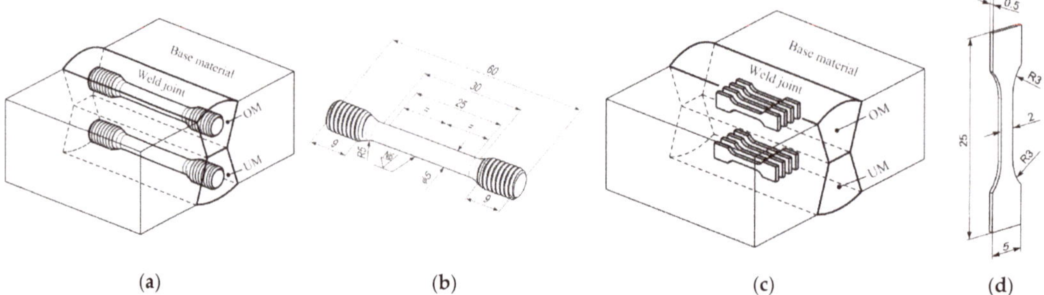

(a)   **(b)**   **(c)**   **(d)**

Figure 2. Positions of tensile specimens in weld joint and geometry of MTSs: (**a**) orientation and position of round tensile specimen in weld metal; (**b**) round specimen geometry; (**c**) orientation and position of set of mini tensile specimens in weld metal; (**d**) mini tensile specimen geometry.

The local mechanical properties for the OM and UM weld material and corresponding HAZ are presented in Figure 3.

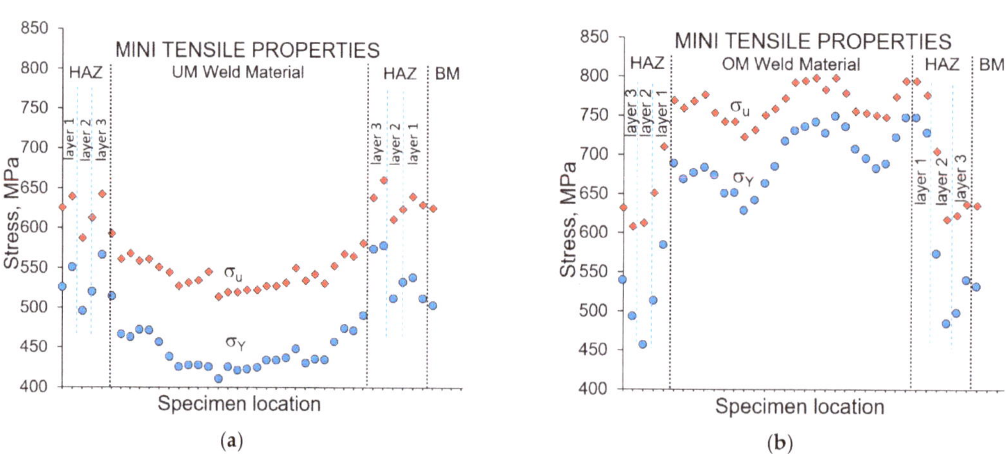

(a)   **(b)**

Figure 3. Results of tensile testing of both welded metals from mini tensile specimens: (**a**) under-match weld joint; (**b**) over-match weld joint.

With the presented sample, a three-point SENB specimen was analysed in accordance with the standard ASTM E1820. The sample thickness was W = 25 mm, and the initial crack length a_0 = 11.2 mm (configuration 1) or 7.9 mm (configuration 2). Figure 4 shows schematic view of specimen orientation in a welded plate and testing manner. The fracture toughness testing was performed at room temperature, 24.5 °C, and under a constant stroke velocity of 0.5 mm/min. The CMOD versus the reaction force F was recorded and compared with the simulation results. The experiments are performed at room temperature (+24 °C) for standard, mini tensile and three point bending specimens by the servo-hydraulic testing machine INSTRON. A single specimen method was used for the crack tip opening displacement (CTOD) testing according to the standard BS 7448 [29]. The CTOD tests were performed under displacement control at a loading rate of 1 mm/min. The load (F), the load point displacement, and the crack mouth opening displacement (CMOD) were recorded. Figure 10a shows the experimentally obtained F vs. CMOD load curve where, after a certain amount of stable crack propagation in the OM metal and after achieving the maximum load, a step of unstable crack propagation is exhibited, followed by stable crack growth in the UM metal.

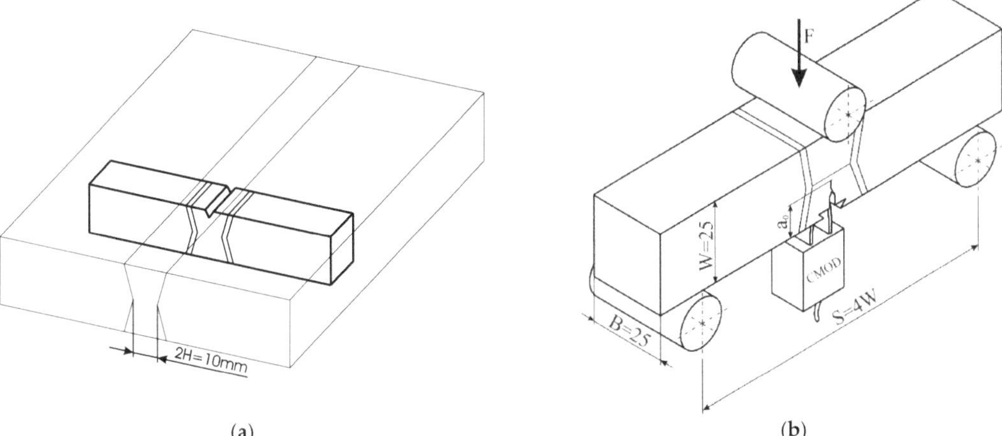

(a)   (b)

Figure 4. Schematic view of specimen orientation in a welded plate and testing manner: (**a**) specimen notch-crack orientation; (**b**) specimen for three point bending fracture toughness testing.

3. Finite Element Simulation

A two-dimensional model of a three-point SENB specimen was modelled according to the testing procedure and specimen geometry, as is shown in Figure 5. All simulations were performed using a commercial finite element method software, SIMULIA Abaqus [30], an implicit dynamic solver with the quasi-static application (quasi-static loading also applied in hardware). The FE model was assembled with the specimen-welded structure and the loading roller (16 mm in diameter) as an analytic rigid body. Displacement over time was defined on the upper roller. The supports on both sides were modelled with the prescribing boundary conditions ($y = 0$) at two nodes; therefore, neighbouring elements had local linear elastic material definition due to stress concentration issues. Therefore, the FE model complexity and computation time were reduced, but the results were not affected by the simplification.

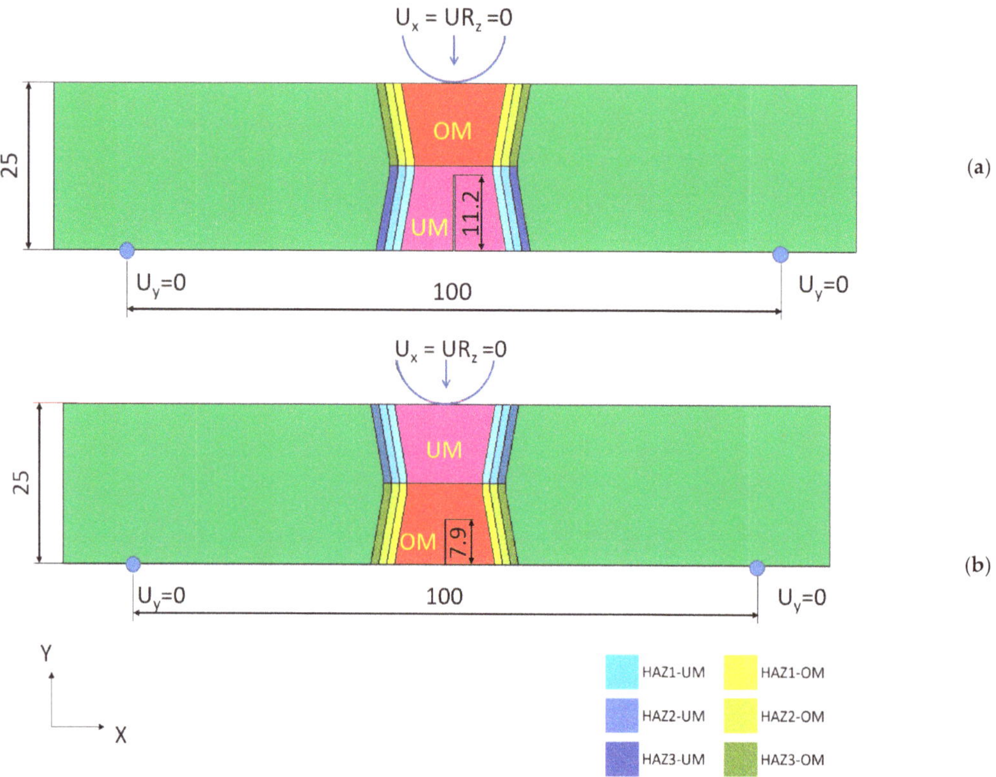

Figure 5. Schematic view on model for FEM with partition for each material enrolled in analysis: (**a**) UM/OM configuration 1 (initial crack in UM); (**b**) OM/UM configuration 2 (initial crack in OM).

The analysed samples are discretised with plane strain finite elements (first order, CPE4 in the region of interest and CPE3 in the not interested regions). A structured mesh with quad elements (size 0.25 mm) was applied between the weld materials and the HAZ. Towards the outside, the element size increased up to 2.0 mm in length, since in this region, the influence on the overall behaviour can be neglected (Figure 6). The FE model consisted of 18,364 finite elements with 18,525 nodes and 18,359 finite elements with 18,525 nodes, as is referred in Table 4, for initial crack in OM and initial crack in UM, respectively. The model thickness was $B = 25$ mm as it was on the tested sample.

The analysed model for configuration 1 consisted of base material, and over- and under-matched material (Figure 5). The initial notch was one finite element wide; the notch length was according to the fatigue pre-crack length in Section 2. Since the mechanical properties within the HAZ change with the distance from the weld interface, three layers of equal thickness of HAZ were defined to describe the material properties' variation in the HAZ. Table 5 shows average material properties.

Table 4. Configuration of both models for FEM analysis.

Sample Configuration	Number of Elements	Number of Nodes
1 (initial crack in UM)	18,350	18,525
2 (initial crack in OM)	18,364	18,525

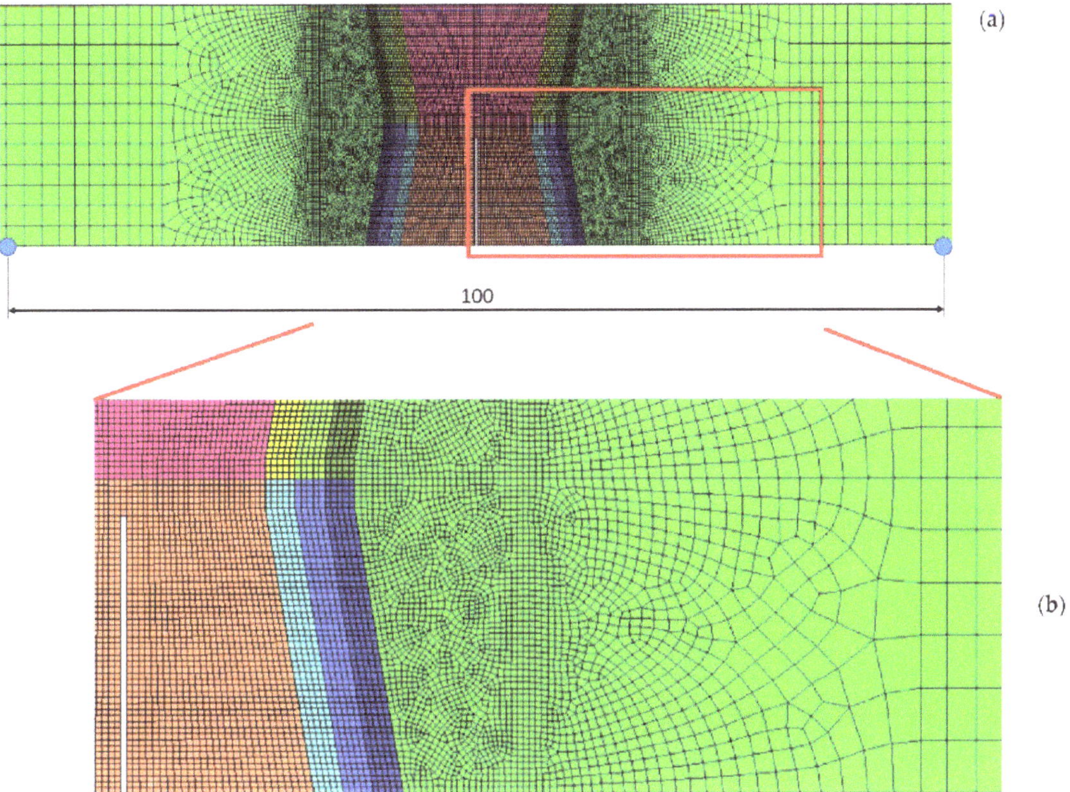

Figure 6. Model for FEM analysis with crack tip in the middle of the weld metal: (**a**) mesh of the entire model; (**b**) detailed mesh in the relevant area.

Table 5. Material parameters for reference BM, OM and UM.

Material Model	E [GPa]	ν [-]	R_{p02} [MPa]	ε_0^{pl} [-]	u_f^{pl} [mm]
Base material	210.0	0.3	530	0.08	0.5
Over matched	210.0	0.3	605	0.08	0.3
Under matched	210.0	0.3	430	0.08	0.3

In addition to the elastic-plastic material model, the ductile damage formation [31] was considered for the BM, OM and UM materials. All material parameters (Young's module, Poisson's ratio, plastic strain hardening curve, plastic strain at damage initiation and critical plastic displacement) were defined according to the uniaxial tensile testing response in Section 2. We used an elastic-plastic ductile damage material model to describe material behaviour because the comparison between curves from tensile experimental testing (full lines) and simulations (dotted lines) showed excellent agreement with each other, and they are plotted in Figure 7. The material damage properties were calibrated for the finite elements size of the tensile specimen model, and the same size was used for the three-point SENB specimen.

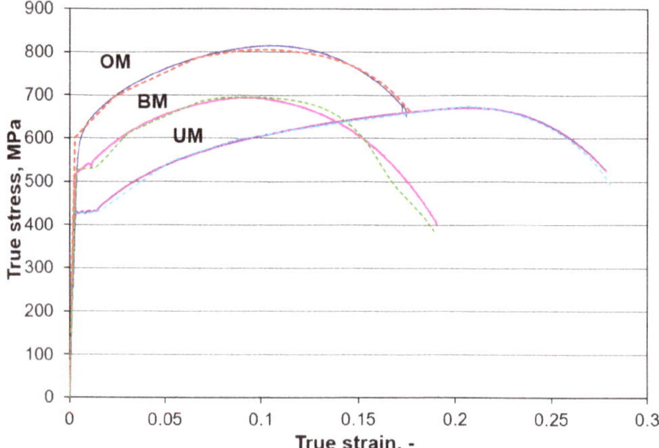

Figure 7. Average mechanical tensile properties obtained by standard tensile testing of round tensile specimens, as is shown in Figure 2a.

The used material models were the base for the material models defined inside the weld and HAZ. As presented above, the global mechanical properties were inhomogeneous in the model, and exhibited a slight local variation inside each material region. Our main goal was to use a modelling technique that includes small local variations in the measured values inside global material regions. The curve shapes were used for given regions and scaled according to the variation in the yield strength Rp02 measured with micro tensile specimens.

We focused on the R_{p02} and R_m values and created five groups for the UM and OM material with respect to how many measured points were inside each group-shares on each level, as seen in Table 6. Figure 8 illustrates the grouping based on the R_{p02} values. Therefore, five material models were created for the UM and OM, and they were based on the reference curve model, as shown in Figure 7, where the true plastic-stress values and damage parameters were scaled according to Table 6. The length interval of each group was proportional to the frequency of properties on the strength level. Further, all elements inside a weld material were assigned randomly to a property, with respect to the shares of each material/group. Figure 9 show the random distribution of elements, with five different mechanical properties for both specimens with different two-filled material properties.

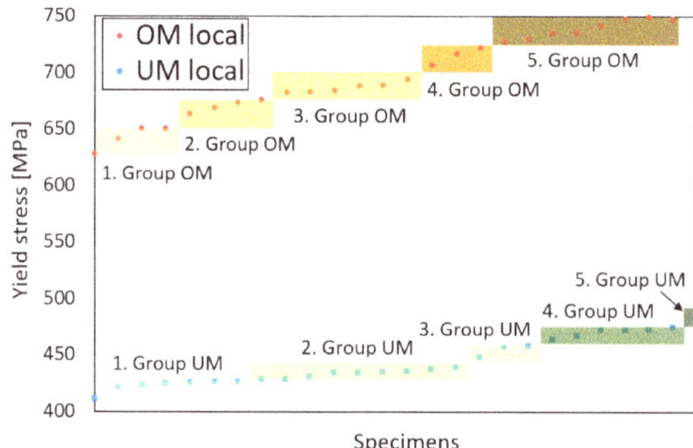

Figure 8. Distribution of yield stress results on 5 equidistant levels for each weld metal.

Table 6. Groups of mechanical properties used in FEM analysis.

	Group	R_{p02} [MPa]	R_m [MPa]	Share
OM weld materials region	1	640.4	730.1	16%
	2	664.6	745.2	16%
	3	688.8	760.3	24%
	4	712.9	775.4	12%
	5	737.1	790.5	32%
UM weld materials region	1	419.5	520.2	27%
	2	435.4	533.7	35%
	3	451.4	547.3	12%
	4	467.3	560.9	23%
	5	483.2	574.4	4%
HAZ at OM side	HAZ1	578	707	
	HAZ2	471	614	
	HAZ3	539	634	
HAZ at UM side	HAZ1	573	647	
	HAZ2	504	599	
	HAZ3	545	640	

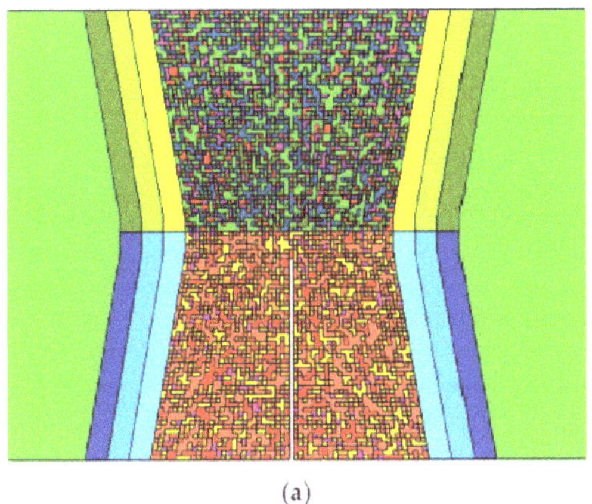

(a)

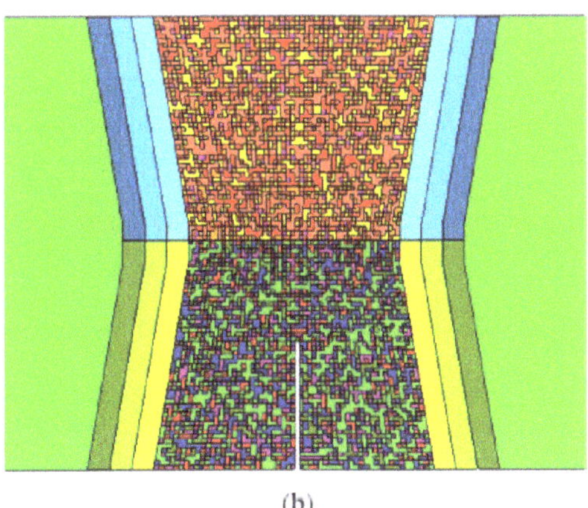

(b)

Figure 9. Random distribution of 5 group of mechanical properties in mesh elements for both combinations of specimens: (**a**) UM/OM configuration 1 (initial crack in UM); (**b**) OM/UM configuration 2 (initial crack in OM).

The HAZ was modelled with three equal thick layers (1.25 mm wide, Figure 9) according to the variation in material properties in the HAZ, see Table 5 and Figure 3, with scaling of the plastic-stress values and damage parameters. The input parameters for the scaling were the averages of the R_{p02} and R_m values from both sides of the weld. An example of the FE-Model with five groups for the UM/OM weld and additional three layers for the HAZ is presented in the Figure 9.

4. Results

The reaction force on the middle loading roller and the CMOD were measured and compared between the experimental and simulation responses for the three-point SENB specimen. As mentioned above, the material properties were assigned randomly to the elements inside each of the two weld regions; therefore, three models were analysed to show the influence of the random distribution/assignment.

Figure 10 shows the reaction force versus CMOD for the two configurations compared with the experimental curves. The behaviour between FE simulation and experimental testing showed a similar response along the whole loading and unloading sequences, as well as some instabilities came out from the simulation, as they appeared during fracture mechanics testing.

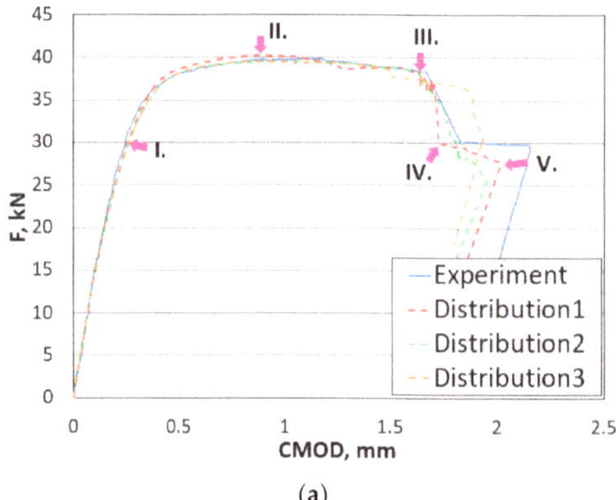
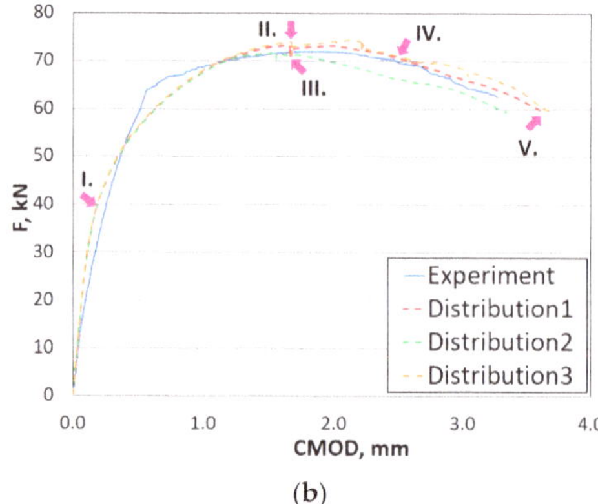

Figure 10. Loading curves for both specimens obtained by experimental measurements and numerical simulations: (**a**) UM/OM configuration 1 (initial crack in UM); (**b**) OM/UM configuration 2 (initial crack in OM).

Figure 10a shows the case when the crack start propagates from the UM along the symmetry line and then bends on the interface between the UM and OM material region and continues along the interface between OM and the neighbouring HAZ region for the UM/OM configuration. Nearly the same observations were seen in the fracture mechanics experiment. The three numerical models (configuration UM/OM) with different material properties distribution, predicted different crack propagation paths due to the slightly different material properties distribution. Nevertheless, the stiffness behaviour predicted by the three random distributions was very similar.

In order to present the stress state at characteristic points of loading F vs. CMOD, for each numerically obtained curve five characteristics points were selected, marked by I.–V. Figure 11a I. Shows that the highest von Mises stress concentration appeared far from the crack tip and behind the fusion line between the UM and OM weld metal. Figure 11a II. shows that the crack path follows the maximum von Mises stress path (left or right) from the crack tip to the HAZ-OM-UM triple point, where the crack turned and propagated between the over-matched weld metal and the HAZ. Figure 11a from III. to VI. shows that the maximum von Mises stress remained at the crack tip between the OM and HAZ fusion line. Figure 11b I. shows that the maximum von Mises stress appeared at the crack tip in the OM weld metal. At the point of maximum sustained loading (Figure 11b) II. a maximum von Mises area appeared in the OM and stable crack growth straight to the fusion line between the OM-UM weld metal, as shown in Figure 11b III.–VI. The numerically obtained results, which were compared with the experimental fracture behaviour of the three-point SENB standard specimens by ASTM E-1820, showed a good agreement on the load vs. crack mouth opening displacement curves, as well as a comparable match between the numerically simulated and metallographic measurement deviation of the crack paths. The simulation results show that the fracture behaviour of a three-point SENB specimen with a crack in the middle of a globally heterogeneous weld with good agreement with the experimental results can be described on the basis of tensile stress-strain curves obtained from standard, and scaled with mini-tensile specimens, for each material microstructure.

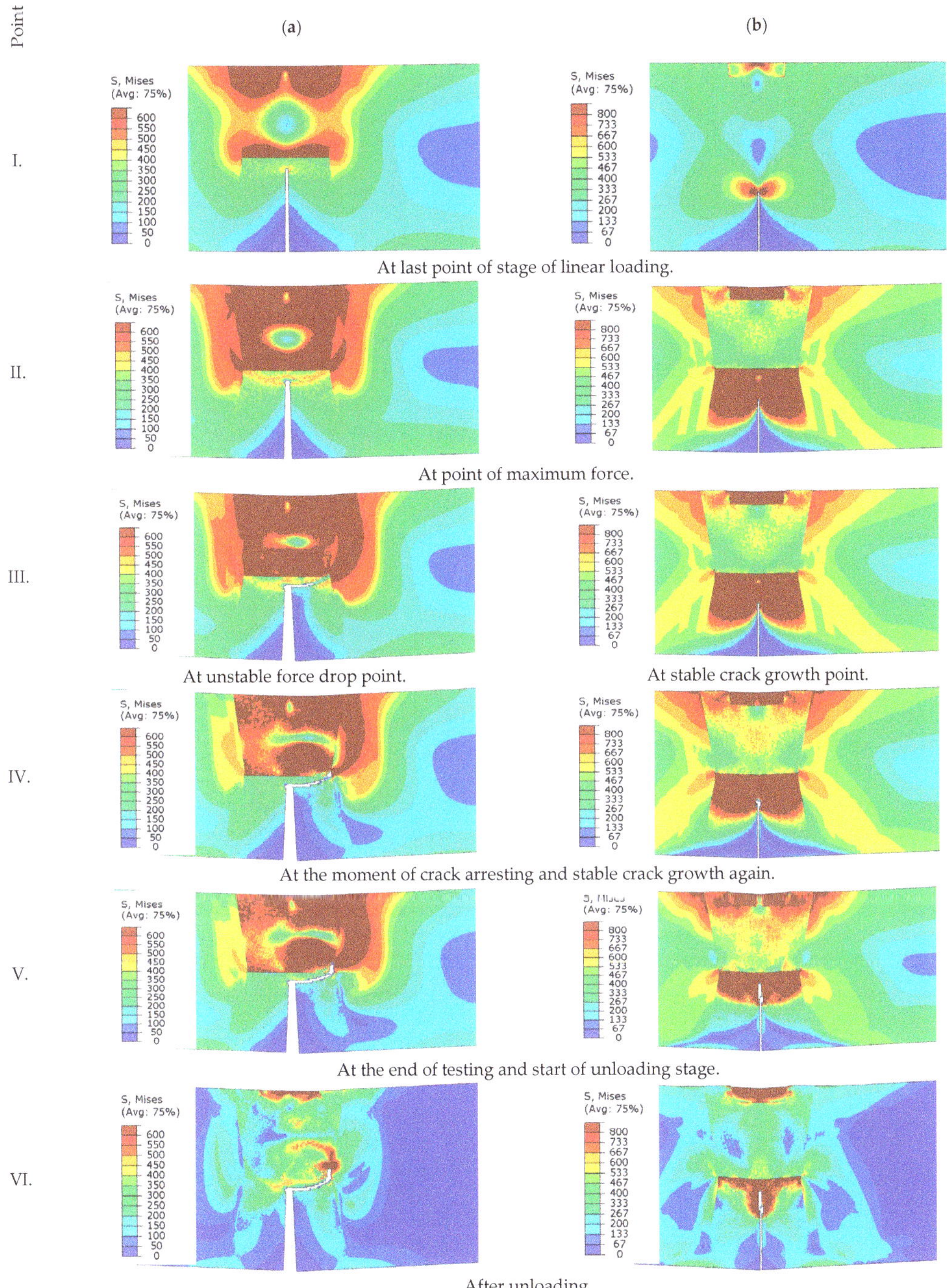

Figure 11. Crack path obtained by FEM simulation of both specimens in characteristic points of loading: (**a**) UM/OM configuration 1 (initial crack in UM); (**b**) OM/UM configuration 2 (initial crack in OM).

Figure 12 shows crack paths for both specimens made by polishing and etching of both halves of the tested specimens.

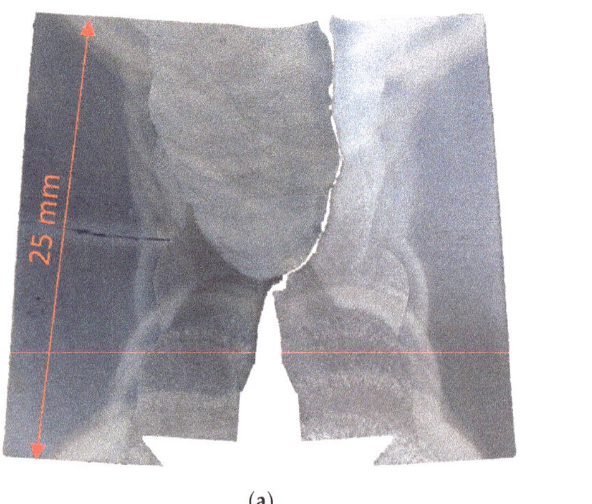

(**a**)

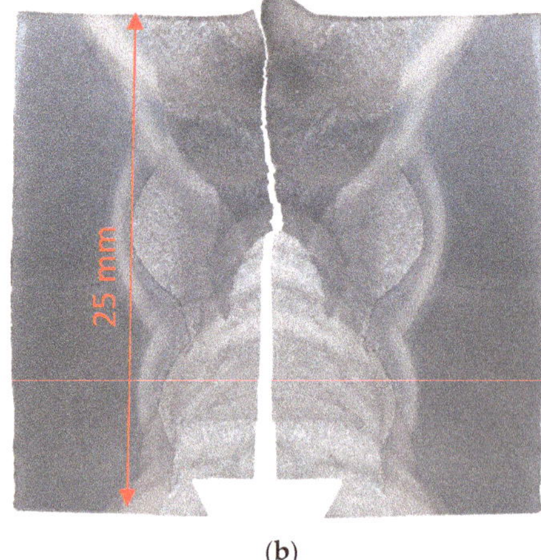

(**b**)

Figure 12. Crack path after fracture toughness testing of both specimens, etched by 4% Nital: (**a**) UM/OM configuration 1 (initial crack in UM); (**b**) OM/UM configuration 2 (initial crack in OM).

5. Conclusions

A numerical investigation was carried out to analyse the effect of local variation in material properties to simulate fracture behaviour in a mismatched X-weld joint. We simulated the fracture behaviour of the strength mismatched multi pass welds numerically successfully by using global and local variations in material properties. The input data were the mechanical material property curves measured from the standard and mini-tensile tests. In the simulations we used the elastic-plastic and ductile damage model in the ABAQUS [31] software by arranging five local areas randomly for varying local properties. The random variation in local material properties where the proportion of one property remains constant did not cause significant deviations between the results of the numerical simulations up to the maximum load for both simulated configurations. The response curves differed after maximum load during the damage process, and the same deviation appeared from the experimentally obtained response. We demonstrated that the response mainly has an effect on the local variation in the properties, which evidently appeared in the multi pass welds. The local crack growth instability phenomena appeared simultaneously in the simulation for configuration 1, as is seen from the experimental curve. We can conclude that, by the using finite element simulation, it is possible to analyse fracture behaviour of the strength mismatched weld in detail by using global and local material properties in the elastic-plastic ductile damage model.

The numerical results also show that the local changes in tensile properties in the microstructure of an individual weld material (either in the OM or the UM) affect the load crack mouth opening curve obtained during fracture mechanical testing of the bending specimen significantly, as well as local instabilities were detected by the simulation. We conclude that, due to the random distribution of locally unequal strength areas, we will always obtain a partially different load curve; nevertheless we can describe the fracture behaviour with very good accuracy globally with the damage model.

The following conclusions appeared in this study:

- The mechanical properties inside a multi pass weld region and HAZ are not constant, and this inhomogeneity should be included in FE simulation. An FE-modelling approach, where different properties inside a weld are distributed randomly, and where the inhomogeneity of the HAZ is included, shows sufficient correlation with the experimental results;

- Similar stiffness responses, reaction force versus CMOD between experimental and simulation were observed. Small changes in crack paths appeared due to the idealised weld geometries;
- FE-modelling with randomly distributed material properties should be considered further, together with simulation of similar welded structures, especially those weld connections that present a "weak-spot" for the entire structure.

Author Contributions: Sample preparation and experimental testing, N.G.; FEM numerical simulations, L.S. and J.P.; supervision, N.G., writing L.S., J.P. and N.G. All authors have read and agreed to the published version of the manuscript.

Funding: Experiments are funding by Marie Curie Fellow Programme No. MCFI-1999-00314.

Institutional Review Board Statement: Not applicable.

Informed Consent Statement: Not applicable.

Data Availability Statement: Not applicable.

Acknowledgments: The authors are grateful for the financial support from the Slovenia Research Agency, Core Research Funding Programme P2-0137. Nenad Gubeljak acknowledges the European Commission in the frame of the Marie Curie Fellow Programme with the project titled »An estimation of the integrity of a welded structure regarding Constraint« No. MCFI-1999-00314 for funding staying at the GKSS Research Centre Geesthacht, today the Helmholtz Zentrum Hereon, Institute of Material and Process Design Germany.

Conflicts of Interest: The authors declare no conflict of interest.

References

1. Qian, Y.; Zhao, J. Fracture Toughness Calculation Method Amendment of the Dissimilar Steel Welded Joint Based on 3D XFEM. *Metals* **2019**, *9*, 509. [CrossRef]
2. Brocks, W.; Cornec, A.; Scheider, I. 3.03—Computational Aspects of Nonlinear Fracture Mechanics. In *Comprehensive Structural Integrity Fracture of Materials from Nano to Macro*; Milne, I., Ritchie, R.O., Karihaloo, B.B.T., Eds.; Pergamon: Oxford, UK, 2003; pp. 127–209, ISBN 978-0-08-043749-1.
3. Kim, S.-H.; Han, J.-J.; Kim, Y.-J. Limit Load Solutions of V-groove Welded Pipes with a Circumferential Crack at the Centre of Weld. *Procedia Mater. Sci.* **2014**, *3*, 706–713. [CrossRef]
4. Kim, Y.-J.; Schwalbe, K.-H. Mismatch effect on plastic yield loads in idealised weldments: II. Heat affected zone cracks. *Eng. Fract. Mech.* **2001**, *68*, 183–199. [CrossRef]
5. Zerbst, U. Application of fracture mechanics to welds with crack origin at the weld toe: A review Part 1: Consequences of inhomogeneous microstructure for materials testing and failure assessment. *Weld. World* **2019**, *63*, 1715–1732. [CrossRef]
6. Wang, H.T.; Wang, G.Z.; Xuan, F.-Z.; Tu, S.-T. An experimental investigation of local fracture resistance and crack growth paths in a dissimilar metal welded joint. *Mater. Des.* **2013**, *44*, 179–189. [CrossRef]
7. Schork, B.; Kucharczyk, P.; Madia, M.; Zerbst, U.; Hensel, J.; Bernhard, J.; Tchuindjang, D.; Kaffenberger, M.; Oechsner, M. The effect of the local and global weld geometry as well as material defects on crack initiation and fatigue strength. *Eng. Fract. Mech.* **2018**, *198*, 103–122. [CrossRef]
8. Zerbst, U.; Madia, M.; Vormwald, M.; Beier, H.T. Fatigue strength and fracture mechanics—A general perspective. *Eng. Fract. Mech.* **2018**, *198*, 2–23. [CrossRef]
9. Zerbst, U.; Ainsworth, R.A.; Beier, H.T.; Pisarski, H.; Zhang, Z.L.; Nikbin, K.; Nitschke-Pagel, T.; Münstermann, S.; Kucharczyk, P.; Klingbeil, D. Review on fracture and crack propagation in weldments—A fracture mechanics perspective. *Eng. Fract. Mech.* **2014**, *132*, 200–276. [CrossRef]
10. Peng, Y.; Wu, C.; Gan, J.; Dong, J. Characterization of heterogeneous constitutive relationship of the welded joint based on the stress-hardness relationship using micro-hardness tests. *Constr. Build. Mater.* **2019**, *202*, 37–45. [CrossRef]
11. Hemer, A.; Milovic, L.; Grbovic, A.; Aleksic, B.; Aleksic, V. Numerical determination and experimental validation of the fracture toughness of welded joints. *Eng. Fail. Anal.* **2020**, *107*, 104220. [CrossRef]
12. Song, Y.; Hua, L. Influence of inhomogeneous constitutive properties of weld materials on formability of tailor welded blanks. *Mater. Sci. Eng. A* **2012**, *552*, 222–229. [CrossRef]
13. Fan, K.; Wang, G.Z.; Tu, S.T.; Xuan, F.Z. Geometry and material constraint effects on fracture resistance behavior of bi-material interfaces. *Int. J. Fract.* **2016**, *201*, 143–155. [CrossRef]

14. Gubeljak, N. Fracture behaviour of specimens with surface notch tip in the heat affected zone (HAZ) of strength mis-matched welded joints. *Int. J. Fract.* **1999**, *100*, 155–167. [CrossRef]
15. Predan, J.; Gubeljak, N.; Kolednik, O.; Fischer, F.D. The change of the local crack driving force in an under-match welded joint. In Proceedings of the 15th European Conference of Fracture, Stockholm, Sweden, 11–13 August 2004; pp. 1–8.
16. Gubeljak, N.; Kolednik, O.; Predan, J.; Oblak, M. Effect of strength of mismatch interface on crack driving force. In Proceedings of the Advanced Fracture damage Mechanics, 3rd International Conference of Fracture and Damage Mechanics—FDM 2003, Paderborn, Germany, 2–4 September 2003; Volume 251–252, pp. 235–244.
17. Predan, J.; Gubeljak, N.; Kolednik, O. On the local variation of the crack driving force in a double mismatched weld. *Eng. Fract. Mech.* **2007**, *74*, 1739–1757. [CrossRef]
18. Gubeljak, N. Determination of lower bound fracture toughness of a high strength low alloy steel welded joint. *Sci. Technol. Weld. Join.* **2005**, *10*, 252–258. [CrossRef]
19. Gubeljak, N.; Legat, J.; Koçak, M. Effect of fracture path on the toughness of weld metal. *Int. J. Fract.* **2002**, *115*, 343–359. [CrossRef]
20. Kozak, D.; Gubeljak, N.; Konjatić, P.; Sertić, J. Yield load solutions of heterogeneous welded joints. *Int. J. Press. Vessel. Pip.* **2009**, *86*, 807–812. [CrossRef]
21. Ran, M.-M.; Sun, F.-F.; Li, G.-Q.; Kanvinde, A.; Wang, Y.-B.; Xiao, R.Y. Experimental study on the behavior of mismatched butt welded joints of high strength steel. *J. Constr. Steel Res.* **2019**, *153*, 196–208. [CrossRef]
22. Wang, W.-K.; Liu, Y.; Guo, Y.; Xu, Z.-Z.; Zhong, J.; Zhang, J.-X. High cycle fatigue and fracture behaviors of CrMoV/NiCrMoV dissimilar rotor welded joint at 280 °C. *Mater. Sci. Eng. A* **2020**, *786*, 139473. [CrossRef]
23. Chen, C.; Chiew, S.-P.; Zhao, M.-S.; Lee, C.-K.; Fung, T.-C. Influence of cooling rate on tensile behaviour of S690Q high strength steel butt joint. *J. Constr. Steel Res.* **2020**, *173*, 106258. [CrossRef]
24. Feng, J.C.; Rathod, D.W.; Roy, M.J.; Francis, J.A.; Guo, W.; Irvine, N.M.; Vasileiou, A.N.; Sun, Y.L.; Smith, M.C.; Li, L. An evaluation of multipass narrow gap laser welding as a candidate process for the manufacture of nuclear pressure vessels. *Int. J. Press. Vessel. Pip.* **2017**, *157*, 43–50. [CrossRef]
25. Long, J.; Zhang, L.-J.; Zhang, Q.-B.; Wang, W.-K.; Zhong, J.; Zhang, J.-X. Microstructural characteristics and low cycle fatigue properties at 230 °C of different weld zone materials from a 100 mm thick dissimilar weld of ultra-supercritical rotor steel. *Int. J. Fatigue* **2020**, *130*, 105248. [CrossRef]
26. Wu, X.; Shuai, J.; Xu, K.; Lv, Z.; Shan, K. Determination of local true stress-strain response of X80 and Q235 girth-welded joints based on digital image correlation and numerical simulation. *Int. J. Press. Vessel. Pip.* **2020**, *188*, 104232. [CrossRef]
27. Ran, M.-M.; Sun, F.-F.; Li, G.-Q.; Wang, Y.-B. Mechanical behaviour of longitudinal lap-welded joints of high strength steel: Experimental and numerical analysis. *Thin-Walled Struct.* **2021**, *159*, 107286. [CrossRef]
28. Ladislav, N.; Mitsuyoshi, T. Creation of Imperfections for Welding Simulations. *Comput. Model. Eng. Sci.* **2011**, *82*, 253–264.
29. *BS 7448-4, Fracture Mechanics Toughness Tests—Part 4: Fracture Mechanics Toughness Tests. Method for Determination of Fracture Resistance Curves and Initiation Values for Stable Crack Extension in Metallic Materials*; British Standard Institution: London, UK, 1991.
30. Abaqus Simulia. Available online: https://www.3ds.com/products-services/simulia/products/abaqus/ (accessed on 2 March 2021).
31. Abaqus Ductile Damage Model. Available online: https://abaqus-docs.mit.edu/2017/English/SIMACAEMATRefMap/simamat-c-damageevolductile.htm (accessed on 12 March 2021).

Article

Numerical and Experimental Investigations of Fracture Behaviour of Welded Joints with Multiple Defects

Mihajlo Aranđelović [1], Simon Sedmak [1], Radomir Jovičić [1], Srđa Perković [2], Zijah Burzić [2], Dorin Radu [3,*] and Zoran Radaković [4]

[1] Innovation Centre of the Faculty of Mechanical Engineering, 11120 Belgrade, Serbia; Mixaylo23@gmail.com (M.A.); simon.sedmak@yahoo.com (S.S.); rjovicic@mas.bg.ac.rs (R.J.)
[2] Military Technical Institute, 11030 Belgrade, Serbia; perkovic.srdja@gmail.com (S.P.); zijah.burzic@vti.vs.rs (Z.B.)
[3] Faculty of Civil Engineering, University of Transylvania, 500036 Brașov, Romania
[4] Faculty of Mechanical Engineering, University of Belgrade, 11120 Belgrade, Serbia; zradakovic@mas.bg.ac.rs
* Correspondence: dorin.radu@unitbv.ro

Abstract: Current standards related to welded joint defects (EN ISO 5817) only consider individual cases (i.e., single defect in a welded joint). The question remains about the behaviour of a welded joint in the simultaneous presence of several different types of defects, so-called multiple defects, which is the topic of this research. The main focus is on defects most commonly encountered in practice, such as linear misalignments, undercuts, incomplete root penetration, and excess weld metal. The welding procedure used in this case was metal active gas welding, a common technique when it comes to welding low-alloy low-carbon steels, including those used for pressure equipment. Different combinations of these defects were deliberately made in welded plates and tested in a standard way on a tensile machine, along with numerical simulations using the finite element method (FEM), based on real geometries. The goal was to predict the behaviour in terms of stress concentrations caused by geometry and affected by multiple defects and material heterogeneity. Numerical and experimental results were in good agreement, but only after some modifications of numerical models. The obtained stress values in the models ranged from noticeably lower than the yield stress of the used materials to slightly higher than it, suggesting that some defect combinations resulted in plastic strain, whereas other models remained in the elastic area. The stress–strain diagram obtained for the first group (misalignment, undercut, and excess root penetration) shows significantly less plasticity. Its yield stress is very close to its ultimate tensile strength, which in turn is noticeably lower compared with the other three groups. This suggests that welded joints with misalignment and incomplete root penetration are indeed the weakest of the four groups either due to the combination of the present defects or perhaps because of an additional unseen internal defect. From the other three diagrams, it can be concluded that the test specimens show very similar behaviour with nearly identical ultimate tensile strengths and considerable plasticity. The diagrams shows the most prominent yielding, with an easily distinguishable difference between the elastic and plastic regions. The diagrams are the most similar, having the same strain of around 9% and with a less obvious yield stress limit.

Keywords: welded joint; finite element method (FEM); multiple defects; stress concentration

Citation: Aranđelović, M.; Sedmak, S.; Jovičić, R.; Perković, S.; Burzić, Z.; Radu, D.; Radaković, Z. Numerical and Experimental Investigations of Fracture Behaviour of Welded Joints with Multiple Defects. *Materials* **2021**, *14*, 4832. https://doi.org/10.3390/ma14174832

Academic Editor: Nicholas Fantuzzi

Received: 29 June 2021
Accepted: 19 August 2021
Published: 25 August 2021

Publisher's Note: MDPI stays neutral with regard to jurisdictional claims in published maps and institutional affiliations.

1. Introduction

Welded joints are of crucial importance for structural integrity due to their crack sensitivity and material heterogeneity [1,2]. For this reason, welded joints are often locations for stress concentration and crack initiation and growth. The fact that welded joints are typically accompanied by defects (to a lesser or greater extent) further emphasises their importance when assessing the integrity of welded structures. The standards EN ISO 5817 and SRPS EN ISO 6520-1 define the acceptability criteria for welded joint defects, but they consider the presence of a single type of defect in a welded joint. Some other procedures, such as the formerly used PD6493 (which was updated into the BS 7910 standard), consider multiple different defects, but only if they are presented as one large single defect [3]. The subject of multiple defects in welded joints was considered by researchers, but to a far lesser extent. Jovičić et al. [2] pointed out the problems that can occur in welded joints due to their geometry, as defects may cause considerable local increase in stress, possibly resulting in crack initiation. In [1], the same authors analysed the effect of multiple defects on fatigue in the existing standard EN ISO 5817, implying that this widely used standard has additional room for improvement. On the other hand, Kozak et al. [4] considered the influence of stresses induced by linear vertical misalignment of cylindrical parts of a pressure vessel, a case that occurs frequently in practice. Other authors, such as Cerit et al. [5], also focused on the numerical analysis of the influence of defects in welded joints, centred on a specific type of defects—undercuts. As can be seen, even if multiple defects are considered, they are actually repeated single defects, treated using a combination of methods previously mentioned, including both experimental and numerical approaches. Initial steps in this direction can be seen in [6].

In terms of defects, the main focus here is on vertical misalignment of plates, along with excess weld metal, incomplete root penetration, and undercuts. Different combinations of these defects (usually three) are introduced along the length of the welded joint for each plate. The material heterogeneity of welded joints was also taken into account when analysing the behaviour of both experimental test specimens and numerical models made using the finite element method. In addition to the influence of welded joint geometry resulting from the presence of multiple defects on the integrity, the effects of different microstructures in a welded joint should be taken into account, as they can affect the direction in which crack propagation and failure take place. It is of great importance to mention that when analysing the heterogeneity of welded joints, all three zones (the parent material, the heat-affected zone, and the weld metal) should be considered in terms of their different mechanical properties resulting from the welding procedure itself. In this case, only the weld metal and the parent material were taken into account, whereas the heat-affected zone will be observed separately in future experiments and numerical analysis once the approach described here is sufficiently developed and verified.

Hence, it can be seen that the ultimate goal of this paper (and the research that it is a part of as a whole) is to unify all factors that influence the behaviour and integrity of welded joints and structures in the presence of multiple defects, with a particular focus on those that occur frequently in practice.

The research presented in this paper involved a number of stages, which will be explained in detail—starting with the welding of plates with specific defect combinations, followed by numerical simulations based on the dimensions measured on the welded plates, after which the tensile tests were performed, with the goal of verifying the obtained numerical results. Once these comparisons were made, it was concluded that additional analyses are necessary for some numerical models and their corresponding specimens in order to improve the existing models so that they would represent the specimen behaviour more realistically.

Compared with other studies [6], the present research is considering a different approach—taking into account defects and a combination of defects, combining experimental and numerical method approaches, thus assessing the structural integrity of the effect of multiple different defects in the welded joint.

2. Materials and Methods

The methods used for these investigations represent a combination of tensile testing of welded joint specimens made of common low-carbon low-alloy steel and numerical simulations of the same specimens. Thus, the main approach was to combine experimental and numerical methods in order to determine and describe the effect of multiple different defects simultaneously present in the welded joint on its structural integrity.

The parent material used in the research is a common structural steel, S235JR steel [6], with the aim of developing the research methodology, which will later be used also for higher-strength steels. Tables 1 and 2 show the chemical composition and mechanical properties of S235JR steel, respectively. The wire VAC 60 was used as a filler material due to its good mechanical properties (Table 3), suitable chemical composition for the purpose of this investigation (Table 4), and availability. This combination of parent and filler materials resulted in significant overmatching since its yield stress was well above that of the parent material. This fact, along with nonstandard welded joint geometry, largely contributed to the results obtained by both numerical and experimental analyses.

Table 1. Chemical composition of S235JR steel [6].

Element	C	Mn	P	S	N	Cu
(%)	0.17	1.4	0.035	0.035	0.12	0.55

Table 2. Mechanical properties of S235JR steel [6].

Material	Yield Stress R_{eH} (MPa)	Tensile Strength R_m (MPa)	Thickness (mm)
S235JR	235	360–510	12

Table 3. Mechanical properties of VAC 60 [7].

Material	Yield Stress R_{eH} (MPa)	Tensile Strength R_m (MPa)	Elongation (%)	Toughnessat $-40\,^{\circ}$C (J)
VAC 60	>410	510–590	>22	>47

Table 4. Chemical composition of VAC 60 [7].

Element	C	Si	Mn	P	S
(%)	0.08	0.9	1.5	<0.025	<0.025

Regarding the preparation of the tensile test specimens, three specimens with a length of 200 mm and a cross section of 25×10 mm^2 were cut out from each group of defects. The welded plates were made of two pieces with 500×200 mm dimensions, which were then welded with one defect combination in the first half and a different combination in the second half. As a result, four different defect combinations were obtained for half a welded plate each. Hence, a total of 12 specimens were subjected to tensile testing in order to obtain more accurate results in accordance with relevant standards.

The welding parameters that were used for the MAG procedure are shown in Tables 5 and 6 for welded joints with and without misalignment. As regards the filler material used, the VAC 60 wire diameter was 1.6 mm.

Table 5. Welding parameters for joints with misalignment.

Layer	Interpass Temperature	Current (A)	Voltage (V)	Welding Speed (mm/s)	Heat Input (kJ/mm)
Root	Below 150	91	18.8	1.7	0.91
Fill 1	Below 150	110	19.4	2.6	0.74
Fill 2	Below 150	120	19.8	1.9	1.12

Table 6. Welding parameters for joints without misalignment.

Layer	Interpass Temperature	Current (A)	Voltage (V)	Welding Speed (mm/s)	Heat Input (kJ/mm)
Root	Below 150	111	19.3	2.2	0.87
Fill 1	Below 150	141	23.9	3.2	0.95
Fill 2	Below 150	150	22	4.1	0.71

The tensile tests in this research were performed at the Military Technical Institute in Belgrade on an Instron tensile test machine (load capacity of 250 kN). The tensile testing equipment that was used are shown in Figure 1. The test specimens were divided into four defect combination groups. As can be seen in the figure, the specimens were not machined, and the original geometry obtained during welding was preserved. The cross-section area of the specimens was 250 mm^2 (25 × 10), whereas the expected tensile strength was around 360 MPa. Based on these values, it was determined that failure of the specimens should be expected at load levels of 90–95 kN, which was confirmed by an experiment. Tensile tests were thoroughly monitored, including the making of images of all four types of specimens, which will be shown at a later point in this paper. As a result, force-displacement diagrams were obtained, which can then be used to determine the stress–strain curves for the needs of numerical simulations.

Figure 1. Instron tensile test machine used for the experiments.

Experiment preparations were performed in accordance with the EN ISO 15614-1:2017 standard (welding specification and qualification), and the specimens were also cut according to this standard's recommendations. With regard to the tensile test experiment, it was performed based on the standard PN-EN ISO 4136:2013 for destructive testing of welded joints in metallic materials, which is closely related to the EN ISO 6892-1:2020 standard, commonly used for such applications.

Finite element methods were used for numerical simulations due to their simplicity, efficiency, and repeatability, as shown in [6–16], where Abaqus 2017 was used in similar analyses. Both elastic and plastic behaviours were defined in these models, for the parent material and weld metal, using data from Tables 1 and 3 as the input data for the simulation of specimen behaviour under tensile loads. When defining plasticity, Abaqus requires the calculation of true stresses and strain based on the values taken from the stress–strain diagram. This approach was used in the numerical models, which will be shown here. Thus, before being used as input data, stress and corresponding strains are converted into their true values using well-known formulas [17]:

$$\sigma_{true} = \sigma(1 + \varepsilon) \tag{1}$$

$$\varepsilon_{true} = ln(1 + \varepsilon) \tag{2}$$

where σ and ε are engineering stress–strain values.

3. Finite Element Method Simulations

In this research, a total of four numerical models were made with combinations of defects, as shown in Figures 2–5. The finite elements used were CPS4R, four-node bilinear plane stress quadrilateral elements, and their number was around 7500 to 8900 for all models except for the one in Figure 5, which had 17,800 finite elements.

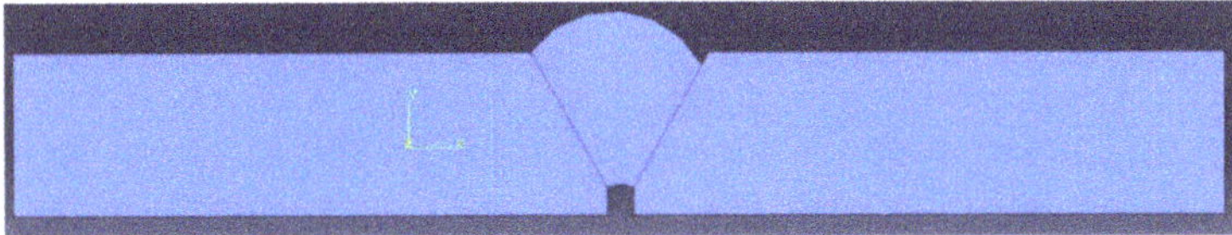

Figure 2. Model with excess weld metal, incomplete root penetration, and undercut.

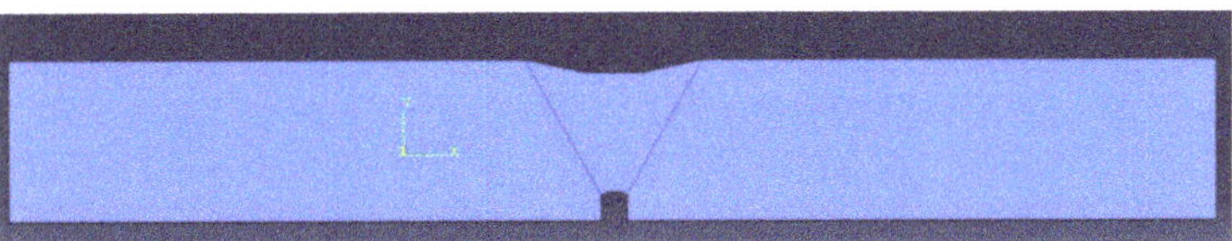

Figure 3. Weld metal sagging and incomplete root penetration.

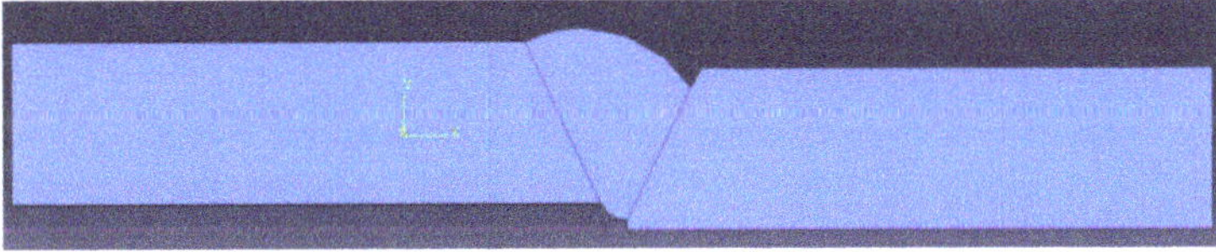

Figure 4. Model with vertical misalignment, incomplete root penetration, and undercut.

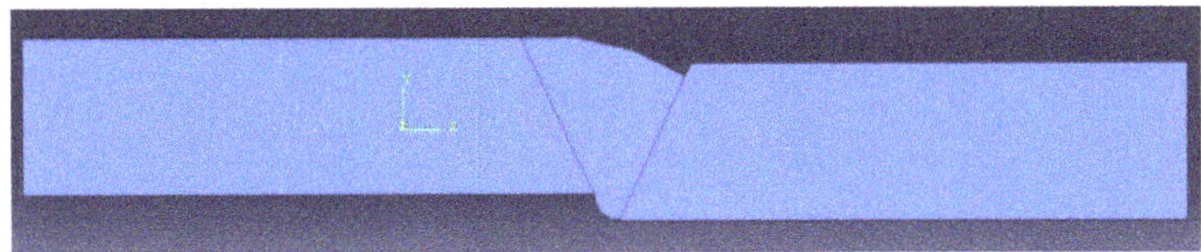

Figure 5. Model with vertical misalignment, undercut, and slight excess root penetration.

The defects represented by the model correspond to the ones in real welded plates in terms of both location and dimensions. Regarding the boundary conditions, one vertical edge of each model is fixed. The load is defined on the opposite edge in the form of tension with a magnitude of 100 MPa. Which side of the model is fixed and which is subjected to tensile loads depends on the experiment—it is based on how the specimen is placed in the tensile test machine since not all specimens are facing the same direction. One advantage of the numerical approach is that boundary conditions and loads can easily switch places if there is a need (e.g., when it turns out that the test specimen is placed in the tensile test machine in the opposite direction). In other words, there is always a possibility that the boundary conditions and loads in the model will be defined in one way, but the real specimen will be placed in the tensile test machine in a way that the tensile load is applied on the end of the specimen that was assumed to be fixed in the model and vice versa. Normally, this would not matter, but the geometry resulting from various defects makes the test specimens asymmetric, which means they will behave in different ways depending on which end is fixed and which is subjected to tension.

The results of each simulation, with the main focus on stress distribution, are shown in Figures 6–9. Stress magnitudes are observed in order to determine the locations of stress concentrations and their relation to the defects. These results were also used as a base in the aforementioned previous research [6]. The comparisons made during that stage of the investigation were the inspiration for the following work in order to explain the somewhat unexpected behaviour that was obtained as a result. As can be seen in Figures 6–9, stress concentrations are indeed highest at the locations of various defects due to the load direction and irregular geometries. In the case of models without misalignment, stress is concentrated in the weld metal, but is still well below the yield stress; thus no plastic strain is induced. Stresses in the parent material remained in the safe (elastic) region. Regarding the models with misalignments, there are two significantly different observations: they had two different locations with noticeably higher stress, and the highest stresses were observed in the parent material, both slightly exceeding the yield stress, which made these cases less favourable despite the overall lower maximal values compared with the first two models. More particularly interesting results were obtained once tensile tests were performed and numerical and experimental results were compared with each other.

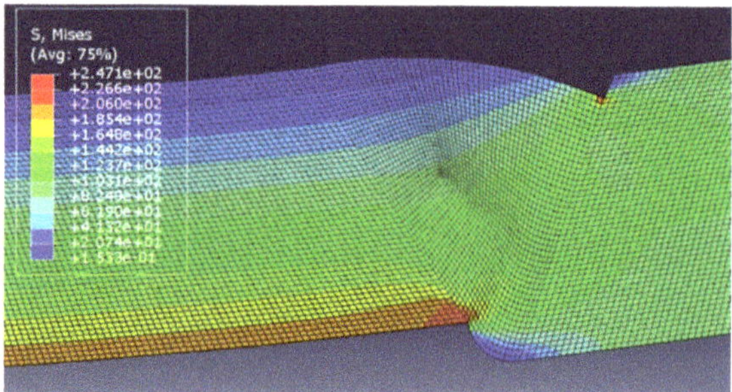

Figure 6. Results for the model with excess weld metal, incomplete root penetration, and undercut.

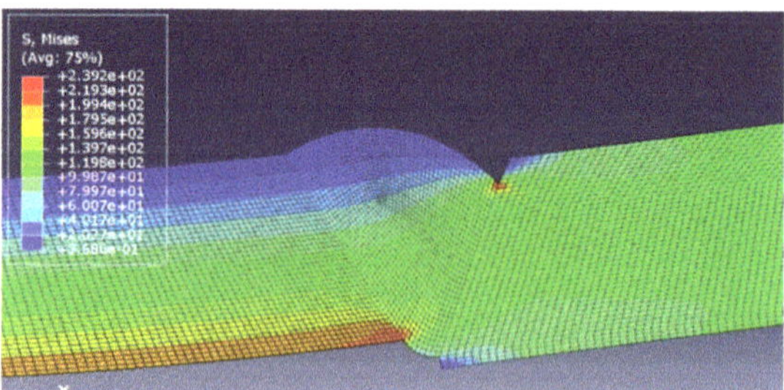

Figure 7. Results for weld metal sagging and incomplete root penetration.

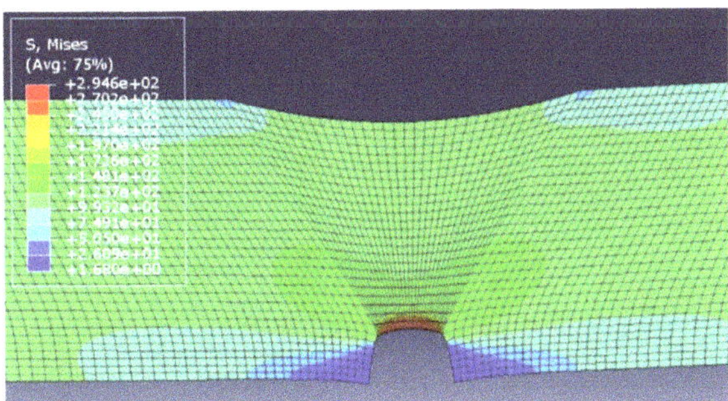

Figure 8. Results for vertical misalignment, incomplete root penetration, and undercut.

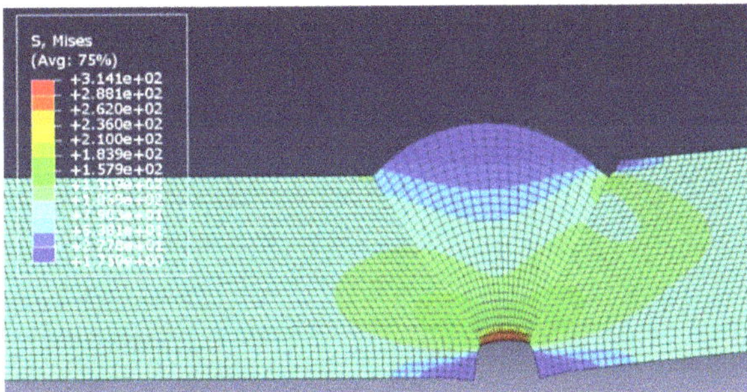

Figure 9. Result for vertical misalignment, undercut, and slight excess root penetration.

4. Experimental Results

It is common practice to verify the results of numerical simulations via experiments, which typically involve numerous types of destructive and nondestructive test methods, as can be seen in [18–29]. The experimental stage that followed numerical simulations involved testing four groups of specimens with each group consisting of three specimens with one of four defect combinations. These specimens, with a cross section of 25 × 10 mm, were cut out of plates with corresponding defect combinations (Figure 10a–d).

(a)   (b)

Figure 10. *Cont.*

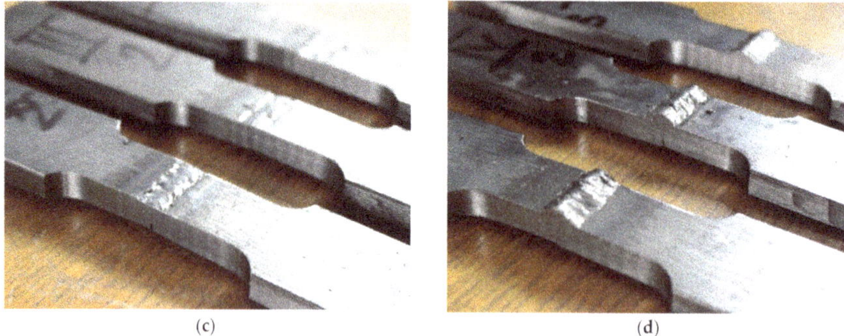

(c)   (d)

Figure 10. Tensile test specimens with different defect combinations: (**a**) group I—misalignment, undercut, slight excess root penetration; (**b**) group II—misalignment, undercut, incomplete root penetration; (**c**) group III—weld face sagging, incomplete root penetration; (**d**) group IV—excess weld metal, undercut, incomplete root penetration.

Figures 11–14 show the tensile test specimens during different stages of the experiment and illustrate how they deformed until fracture for all four groups. This was used for comparison with the obtained numerical results to determine whether the real stress concentration locations (as well as the location where failure occurred) correspond to the ones obtained by the numerical models.

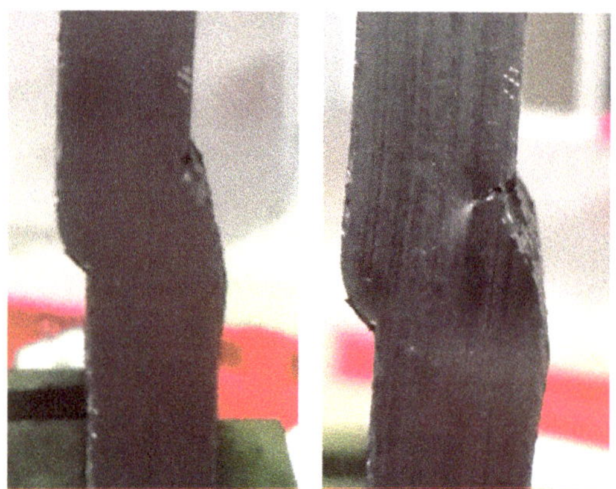

Figure 11. Tensile test specimens during the experiment (misalignment model with slight excess root penetration and undercut) from the initial state to fracture.

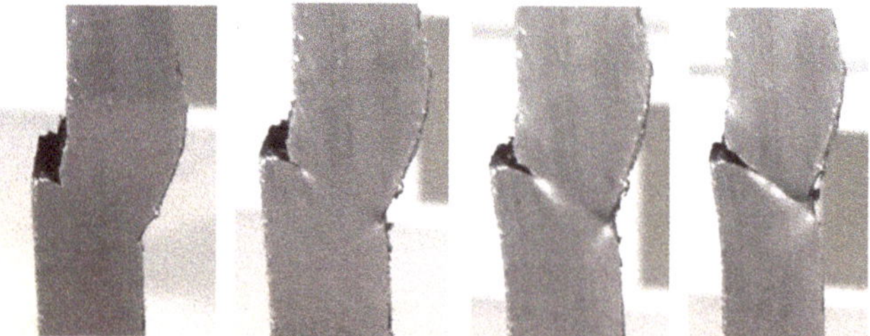

Figure 12. Tensile test specimens during the experiment (misalignment model with incomplete root penetration and undercut) from the initial state to fracture.

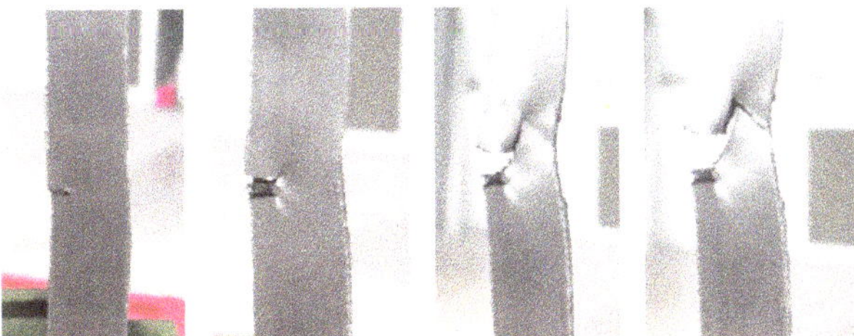

Figure 13. Tensile test specimens during the experiment (weld metal sagging and incomplete root penetration) from the initial state to fracture.

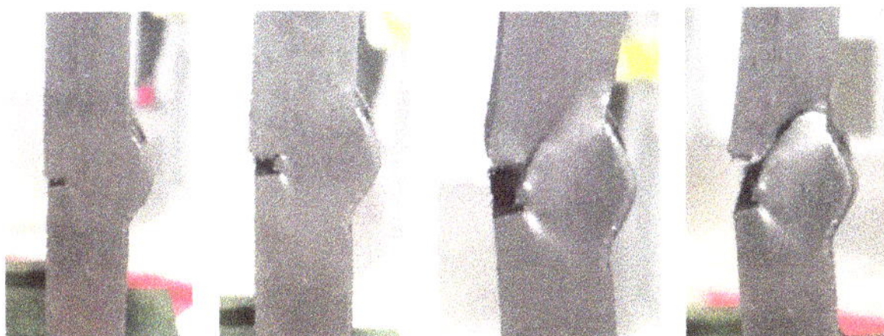

Figure 14. Tensile test specimens during the experiment (excess weld metal, undercut, incomplete root penetration) from the initial state to fracture.

5. Discussion of the Initial Results

Figures 7 and 12 (numerical and experimental results for the misalignment model with incomplete root penetration and undercut), as well as Figures 9 and 14 (numerical and experimental results for the model with excess weld metal, undercut, and incomplete root penetration), indicate only partial agreement between the numerical simulations and the actual experiment. Stress concentrations in both cases are similar and quite high in the undercut (weld face) but different in the root. As can be clearly seen in Figure 12, a crack initiated in the lower part of the root in the tensile test but was on the opposite side of the root (near the 'higher' plate) in the numerical model (Figure 7). In both cases, failure would initiate in the root, although at different locations. The way in which the tensile test specimen ultimately failed indicates a significant plastic strain occurring in the undercut after a certain crack length is reached in the root side. The same is implied by the numerical model, wherein stresses in the undercut are only slightly lower than the main stress concentration in the weld root region. Several possible explanations for this phenomenon are considered: the presence of additional internal defects caused by intentionally poor welding; the need for a more detailed information about the heat-affected zone, which is not considered in this research; and the possibility of modelling the same case with an initial crack in the weld root zone, in order to see how this would affect the stress distribution and overall deformation of the model.

Without the misalignment, however, models and experiments have shown very good agreement in terms of deformation. Test specimens have shown crack growth along the fusion line, and the same would have happened in the models if the applied load had been sufficient to cause the parent material to yield (corresponding to the fusion line in this case). This could occur on either side of the specimen since it is the only perfectly symmetrical one. Coincidentally, this did occur during the experiment—one specimen failed via crack initiation and growth on the root side of the fusion line, opposite of the end of the applied force.

Figure 12 illustrates the most interesting result obtained experimentally. Naturally, the numerical model did not indicate this type of failure, and this result presents a unique challenge on its own—explaining how and why the crack suddenly took an almost 90° turn and went from the parent material/heat-affected zone to the much stronger weld metal.

The answer will be sought via metallography and fractography tests, which are planned as the next stage of this research.

Finally, the stress–strain diagrams obtained for each of the four groups are shown in Figures 14–17. The first diagram, Figure 15, shows significantly less plasticity (especially taking into account that the whole diagram is 'displaced' relative to the origin for reasons related to the force-displacement results; this issue will be properly addressed in the next stages, when higher-quality materials are tested using this methodology). Additionally, the yield stress is very close to the ultimate tensile strength, which in turn is noticeably lower compared with the other three groups. This suggests that welded joints with misalignment and incomplete root penetration are indeed the weakest of the four, either due to the combination of the present defects, or perhaps because of an additional unseen internal defect.

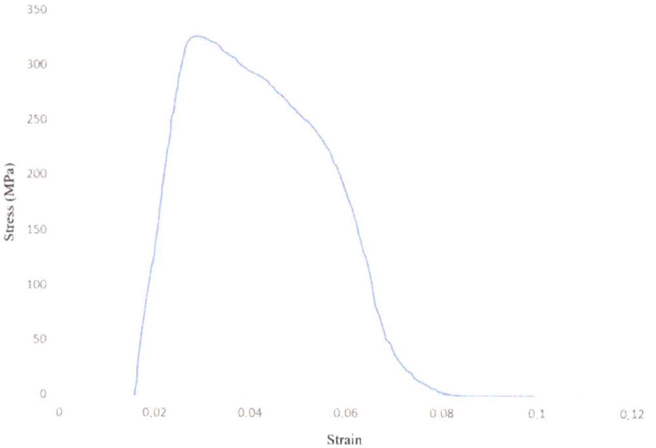

Figure 15. Stress–strain diagram for the group with misalignment, undercut, and excess root penetration (group I).

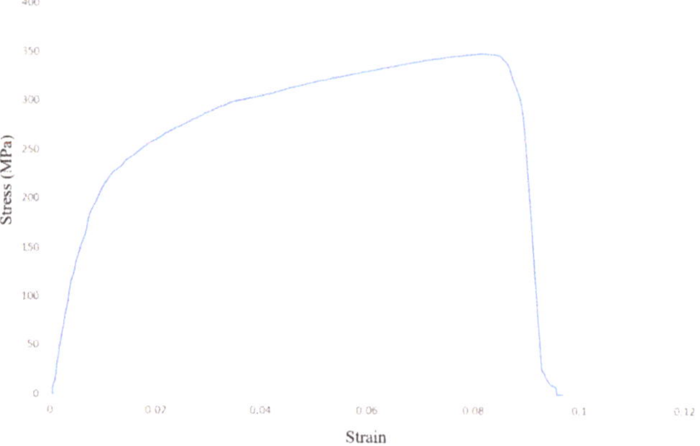

Figure 16. Stress–strain diagram for the group with misalignment, undercut, and incomplete root penetration (group II).

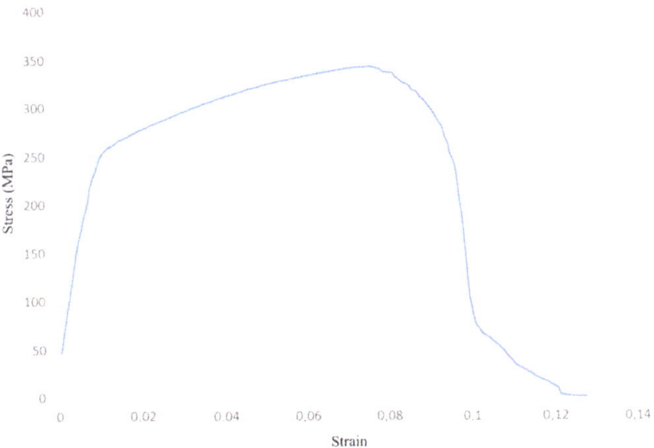

Figure 17. Stress–strain diagram for the group with WM sagging and incomplete root penetration (group III).

From the other three diagrams, Figures 15–17, it can be concluded that the test specimens showed very similar behaviour, with nearly identical ultimate tensile strengths and considerable plasticity. The diagram in Figure 18 shows the most prominent yielding, with an easily distinguishable difference between the elastic and plastic regions. The diagrams in Figures 18 and 19 are the most similar, having the same strain of around 9% and with a less obvious yield stress limit. The aforementioned diagram in Figure 17 has a somewhat higher strain, almost 12%. Another interesting aspect related to Figures 18 and 19 is that they represent specimens from two very different groups—with and without misalignment.

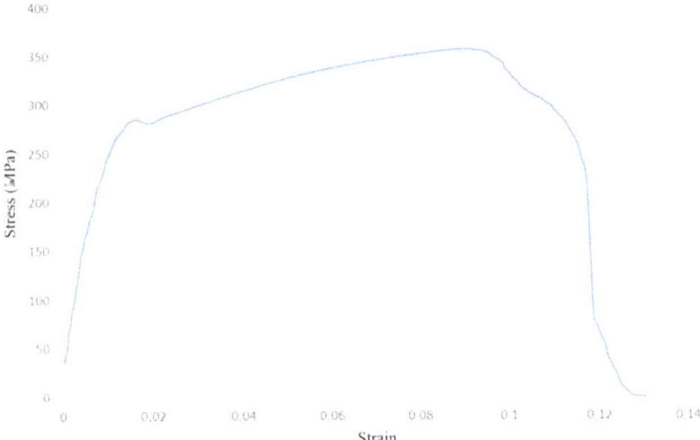

Figure 18. Stress–strain curve for the group with excess WM, undercut, incomplete root penetration (group IV).

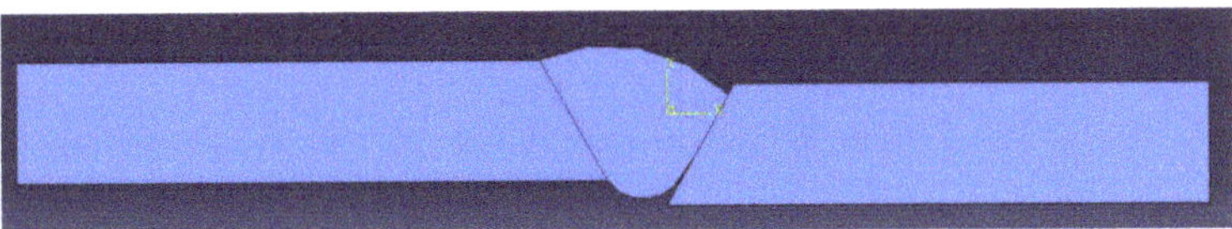

Figure 19. New numerical model geometry, including lack of fusion on the lower side of the welded joint.

6. New Version of the Group 2 Specimen Model

Once the experimental approach confirms that there are noticeable differences in the numerical model and the real specimen, the next step is to include this newly obtained information into the simulation. For this purpose, the depth of lack of fusion is measured and determined to be 4 mm (average value through specimen width). This lack of fusion is then introduced into the numerical model, as shown in Figure 20. It should be noted that this model is the same as the one previously presented in terms of mechanical properties and boundary conditions, and the only changes to its geometry are the added lack of fusion and an increase in gap width, with the second change not having any significant influence on the obtained results.

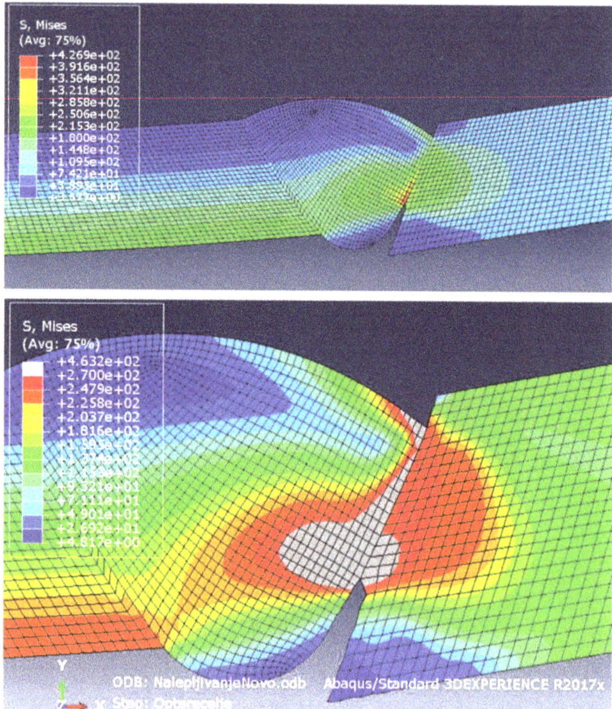

Figure 20. Stress distribution with a tensile load of 100 MPa (**top**) and 120 MPa (**bottom**).

The results obtained for this new version of the model are shown in Figure 20. It can be clearly seen that there are considerable differences compared with the previous model in terms of both stress concentration locations and stress magnitude. The reason the maximum stress in this case is much higher is that the yield stress of the parent material was exceeded significantly, with stresses reaching as much as 270 MPa. This caused additional deformation in the weld metal, with stresses reaching 427 MPa, very close to filler material yield stress (unlike in the first case, where these values are much lower due to stress concentrators not present in the modified version).

At first glance, the second stress concentration located in the undercut seems negligible, and stresses at this location are insufficient to cause plastic strain. However, as can be seen in the bottom image in Figure 20, the situation changes once the load increases. These results are obtained after the tensile load increases to 120 MPa in order to determine the behaviour of the model under a higher plastic strain. At this point, the regions around the undercut also show stress values above the yield stress of the parent material. It should be noted that some results appear grey since the limit for the maximum stress shown in the image is lowered to 270 MPa. This is done in order to obtain a more detailed stress distribution in the PM region since the stresses in the overmatched weld metal are much higher. At this point, the maximum stresses in the WM reached 463 MPa, slightly above the weld metal yield stress, resulting in plasticity in both regions of the welded joint.

7. Conclusions

The aim of this paper is to improve the existing numerical model of a welded joint with a number of surface defects based on real welded joint specimens. The problems with the initial model occur after specimens are experimentally tested, and certain differences are observed between the experimental and numerical results. It is decided to perform additional destructive tests in order to determine whether there are additional factors that contribute to this difference. It is assumed that the difference is influenced by the lack of fusion on the root side of the weld, confirmed by analysing the fractured surface.

- The next stage involves optimisation of the model by introducing newly found defects into the geometry. The obtained results show considerable improvement since the numerical model now behaves in a nearly identical manner as the real specimens in the experiment.
- Stress concentration locations are the same in the experiment and in the new and improved model, and the model itself deforms in a way closely resembling real specimens—with the crack initiating in the weld root—whereas the undercut is subjected to a significant plastic strain, which becomes more prominent with load increase.
- After reaching plasticity in both weld metal and parent material, the behaviour of the numerical model and the experiment show even better mutual agreement.
- In addition to the improved numerical model, based on information obtained on the behaviour of welded joint specimens following the experiment, this model also provides insight into how exactly various defects affect welded joint integrity.
- Incomplete root penetration, combined with lack of fusion, results in a significant adverse effect, whereas the presence of excessive weld metal is completely irrelevant. The undercut is also not of great importance initially, being subjected to compressive stresses due to the combination of load direction and geometry. These stresses only start affecting the welded joint after a significant plastic strain, but at this point, it is already 'too late' for the sake of integrity since the welded joint load-bearing cross section has already significantly decreased, causing failure.

This research results in a methodology that can be used in analysing and solving problems previously not considered for welding defects. With a proper combination of experimental and numerical methods, it is possible to accurately describe the behaviour of a welded joint in the presence of several different defects. Future research, in the aforementioned doctoral thesis, will include the application of this method to welded joints of different materials in order to determine how various defect combinations affect integrity since different types of steel can have rather various mechanical properties and resistance to crack growth and so forth.

For this reason, later experiments that will follow this research will also:

- Consider the mechanical properties of the heat-affected zone of the welded joint, in addition to the already-included properties of the parent material and weld metal, which were both used as input parameters for the performed numerical simulations.
- Focus on a more detailed numerical simulation of the heat-affected zone in welded joints. This further emphasises the importance of the experimental part of the research, which will be taken into account in future research, which will also include determining the heat-affected zone mechanical properties via Vickers hardness tests.

Author Contributions: Conceptualisation—M.A., S.S., Z.B., methodology—Z.B., R.J., S.S.; software—S.S.; experimental validation—M.A., Z.B., S.P., R.J., D.R.; writing—original draft preparation—M.A., S.S.; writing—review and editing—D.R., Z.R. All authors have read and agreed to the published version of the manuscript.

Funding: This work was supported by the Ministry of Education, Science, and Technological Development of the Republic of Serbia (Contract Nos. 451-03-68/2020-14/200135 and 451-03-68/2020-14/200213).

Institutional Review Board Statement: Not applicable.

Informed Consent Statement: Not applicable.

Conflicts of Interest: The authors declare no conflict of interest.

References

1. Sedmak, S.; Jovičić, R.; Sedmak, A.; Aranđelović, M.; Đorđević, B. Influence of multiple defects in welded joints subjected to fatigue loading according to sist EN ISO 5817:2014. *Struct. Integr. Life* **2018**, *18*, 77–81.
2. Jovičić, R.; Sedmak, S.; Tatić, U.; Lukić, U.; Walid, M. Stress state around imperfections in welded joints. *Struct. Integr. Life* **2015**, *15*, 27–29.
3. British Standards Institution. *Guidance on Some Methods for the Derivation of Acceptance Levels for Defects in Fusion Welded Joints*; BSI, Standard PD 6493 from 1980; BSI: London, UK, 1980.
4. Dražen, K.; Pejo, K.; Franjo, M.; Darko, D. Weld misalignment influence on the structural integrity of cylindrical pressure vessel. *Struct. Integr. Life* **2010**, *10*, 153–159.
5. Cerit, M.; Kokumer, O.; Genel, K. Stress concentration effect of undercut defect and reinforcement metal in butt welded joint. *Eng. Fail. Anal.* **2010**, *17*, 571–578. [CrossRef]
6. Aranđelović, M.; Sedmak, S.; Jovičić, R.; Perković, S.; Burzić, Z.; Đorđević, B.; Radaković, Z. Numerical simulation of welded joint with multiple various defects. *Struct. Integr. Life* **2021**, *21*, 103–107.
7. Jeremić, L.; Sedmak, A.; Petrovski, B.; Đorđević, B.; Sedmak, S. Structural integrity assessment of welded pipeline designed with reduced safety. *Tech. Gaz.* **2020**, *27*, 1461–1466. [CrossRef]
8. Sedmak, A. Computational fracture mechanics: An overview from early efforts to recent achievements. *Fatigue Fract. Eng. Mater Struct.* **2018**, *41*, 2438–2474. [CrossRef]
9. Kirin, S.; Sedmak, A.; Zaidi, R.; Grbović, A.; Šarkočević, Ž. Comparison of experimental, numerical and analytical risk assessment of oil drilling rig welded pipe based on fracture mechanics parameters. *Eng. Fail. Anal.* **2020**, *114*. [CrossRef]
10. Berković, M.; Maksimović, S.; Sedmak, A. Analysis of welded joints by applying the finite element method. *Struct. Integr. Life* **2004**, *4*, 75–83.
11. Coshnova, D. The structure of stream and temperature distribution in the reaction shaft of the flash smelting furnace—computer simulation. *Struct. Integr. Life* **2010**, *10*, 129–133.
12. Milovanović, A.; Sedmak, A.; Čolić, K.; Tatić, U.; Đorđević, B. Numerical analysis of stress distribution in total hip replacement implant. *Struct. Integr. Life* **2017**, *17*, 139–144.
13. Đorđević, B.; Sedmak, S.; Tanasković, D.; Gajić, M.; Vučetić, F. Failure analysis and numerical simulation of slab carrying clamps. *Frat. Ed Integrita Struturale* **2021**, *55*, 336–344. [CrossRef]
14. Grbović, A.; Kastratović, G.; Sedmak, A.; Eldwaib, K.; Kirin, S. Determination of optimum wing spar cross section for maximum fatigue life. *Int. J. Fatigue* **2019**, *127*, 305–311. [CrossRef]
15. Chang, P.H.; Teng, T.L. Numerical and experimental investigations on the residual stresses of the butt-welded joints. *Comput. Mater. Sci.* **2004**, *29*, 511–522. [CrossRef]
16. Cho, J.R.; Lee, B.Y.; Moon, Y.H.; van Tyne, C.J. Investigation of residual stress and post weld heat treatment of multi-pass welds by finite element method and experiments. *J. Mater. Process. Technol.* **2004**, *155*, 1690–1695. [CrossRef]
17. Available online: https://info.simuleon.com/blog/converting-engineering-stress-strain-to-true-stress-strain-in-abaqus (accessed on 15 May 2021).
18. Sedmak, S.; Burzić, Z.; Perković, S.; Jovičić, R.; Aranđelović, M.; Radović, L.; Ilić, N. Influence of welded joint microstructures on fatigue behaviour of specimens with a notch in the heat affected zone. *Eng. Fail. Anal.* **2019**, *106*. [CrossRef]
19. Somodi, B.; Kövesdi, B. Residual stress measurements on welded square box sections using steel grades of S235–S960. *Thin-Walled Struct.* **2018**, *123*, 142–154. [CrossRef]
20. Correia, J.A.F.O.; da Silva, A.L.L.; Xin, H.; Lesiuk, G.; Zhue, S.-P.; de Jesus, A.M.P.; Fernandes, A.A. Fatigue performance prediction of S235 base steel plates in the riveted connections. *Structures* **2021**, *30*, 745–755. [CrossRef]
21. Merchant, S.Y. Investigation of Effect of heat input on cooling rate and mechanical property (hardness) of mild steel weld joint by MMAW Process. *Int. J. Mod. Eng. Res.* **2015**, *5*, 34–41.
22. Bosnjak, M.A.S.; Sedmak, S.; Sarkocevic, Z.; Savic, Z.; Radu, D. Determination of damage and repair methodology for the runner manhole of Kaplan turbine at the hydro power plant 'Djerdap 1'. *Struct. Integr. Life* **2016**, *16*, 149–153.
23. Malița, M.; Bolduș, D.; Radu, D.; Băncilă, R. An efficient solution for road steel truss girder welded bridges. *Int. Multidiscip. Sci. GeoConference Surv. Geo–Logy Min. Ecol. Manag.* **2015**, *5*, 37–44.
24. Sedmak, S.; Aranđelović, M.; Jovičić, R.; Radu, D.; Čamagić, I. Influence of Cooling Time t8/5 on Impact Toughness of P460NL1 Steel Welded Joints. *Adv. Mater. Res.* **2020**, *1157*, 154–160. [CrossRef]

25. Djordjevic, B.; Sedmak, A.; Petrovski, B.; Dimic, A. Probability distribution on cleavage fracture in function of Jc for reactor ferritic steel in transition temperature region. *Eng. Fail. Anal.* **2021**, *125*, 105392. [CrossRef]
26. Younise, B.; Rakin, M.; Gubeljak, N.; Medjo, B.; Sedmak, A. Effect of material heterogeneity and constraint conditions on ductile fracture resistance of welded joint zones—Micromechanical assessment. *Eng. Fail. Anal.* **2017**, *82*, 435–445. [CrossRef]
27. Jovičić, R.; Sedmak, A.; Prokić-Cvetković, R.; Bulatović, S.; Buyukyildirim, G. Cracking resistance of AlMg4.5Mn alloy TIG welded joints. *Struct. Integr. Life* **2011**, *11*, 205–208.
28. Fernandino, D.O.; Tenaglia, N.; Boeri, R.E. Microstructural damage evaluation of ferritic-ausferritic spheroidal graphite cast iron. *Frat. Ed Integrita Strutturale* **2020**, *51*, 477–485. [CrossRef]
29. Tomków, J.; Janeczek, A.; Rogalski, G.; Wolski, A. Underwater Local Cavity Welding of S460N Steel. *Materials* **2020**, *13*, 5535. [CrossRef] [PubMed]

 *materials*

Article

Estimation of C^* Integral for Mismatched Welded Compact Tension Specimen

Marko Katinić *, Dorian Turk, Pejo Konjatić and Dražan Kozak

Mechanical Engineering Faculty, University of Slavonski Brod, Trg Ivane Brlic Mazuranic 2,
35000 Slavonski Brod, Croatia; dturk@unisb.hr (D.T.); pkonjatic@unisb.hr (P.K.); dkozak@unisb.hr (D.K.)
* Correspondence: mkatinic@unisb.hr

Abstract: The C^* integral for the compact tension (CT) specimen is calculated using the estimation equation in ASTM E1457-15. This equation was developed based on the assumption of material homogeneity and is not applicable to a welded CT specimen. In this paper, a modified equation for estimating the C^* integral for a welded compact tension (CT) specimen under creep conditions is proposed. The proposed equation is defined on the basis of systematically conducted extensive finite element (FE) analyses using the ABAQUS program. A crack in the welded CT specimen is located in the center of the heat-affected zone (HAZ), because the most severe type IV cracks are located in the HAZ. The results obtained by the analysis show that the equation for estimating the C^* integral in ASTM E1457-15 can underestimate the value of the C^* integral for creep-soft HAZ and overestimate for creep-hard HAZ. Therefore, the proposed modified equation is suitable for describing the creep crack growth (CCG) of welded specimens.

Keywords: creep; C^* integral; mismatched weld; CT specimen; finite element analysis

Citation: Katinić, M.; Turk, D.; Konjatić, P.; Kozak, D. Estimation of C^* Integral for Mismatched Welded Compact Tension Specimen. *Materials* **2021**, *14*, 7491. https://doi.org/10.3390/ma14247491

Academic Editor: Antonino Squillace

Received: 19 October 2021
Accepted: 30 November 2021
Published: 7 December 2021

Publisher's Note: MDPI stays neutral with regard to jurisdictional claims in published maps and institutional affiliations.

1. Introduction

Welding technology is one of the most common ways of joining metals in the chemical and petrochemical industries. However, welds may contain imperfections such as lack of fusion, under-cuts, porosity, and slag inclusion of certain height and length at certain locations in the weld [1,2]. Such irregularities in welds usually act as crack initiation sites. The failure of any of the welded joints in the process industry is uncomfortable at best and can lead to catastrophic accidents at worst. For safe and reliable operation of welded structures, periodic non-destructive testing and appropriate assessment of structural integrity should be performed.

There are increasing demands for reducing CO_2 emissions in the chemical and petrochemical industries and in thermal power plants. Increasing the operating temperatures of the plant increases the energy efficiency, thus reducing CO_2 emissions. Higher operating temperatures and higher stresses increase the risk of welded structure failure due to creep conditions. It is therefore necessary to develop a method to predict the CCG in welded joints [3].

It is most likely that failure due to creep of high-temperature components begins in welded joints. The most severe cracks in welded joints are type IV cracks in the HAZ. These cracks occur due to multi-axial stresses caused by the material constraint effect between the welding constituents and large creep strain of the soft HAZ [4]. Many previous studies [5–8] have shown that material constraint plays an important role in CCG in welded joints. The material constraint is due to the mismatch effect of creep properties in the welded joint components. There are a number of experimental and numerical studies on the influence of the mismatch effect on the fracture mechanics parameter C^* and the CCG rate in welded joints [9–14].

Many studies have shown that in addition to material constraint, the CCG behavior of welded joints is also influenced by geometric constraint [15–17]. Li et al. [18] investigated the effect of the interaction of material constraint and geometry constraint on the CCG rate in welded joints. A FE method based on the ductility exhaustion model was used in this study. A FE method was applied to two specimen types (the compact tension and middle crack tension specimen) with different material mismatches. The research established the existence of the effect of the interaction between material and geometry constraints on the CCG rate. Under the condition of low geometry constraint, the effects of material constraint on the CCG rate become more obvious [18].

Creep crack growth under widespread steady-state creep conditions is characterized by the C^* integral. The C^* integral for the compact tension (CT) specimen is calculated using the estimation equation in ASTM E1457-15 [19]. However, this equation was developed based on the assumption of material homogeneity. As the above studies have shown, the influence of material and geometry constraints on C^* and the CCG rate of welded joints cannot be ignored. Therefore, the C^* estimation equation according to ASTM E1457-15 needs to be modified for application to welded CT specimens. Xuan et al. [3] proposed a model of equivalent material and developed a modified equation for the C^* integral applying limit load solutions. However, the HAZ was not taken into consideration in this study.

In this paper, extensive elastic-creep two-dimensional finite element analysis was performed for welded CT samples with different crack lengths located in the HAZ. The general purpose finite element program ABAQUS [20] was used to calculate the C^* integral. A systematic parametric study of the dependence of the C^* integral on the mismatch of material properties, HAZ width and crack length was performed. Based on the performed analyses, a modified equation for estimating the C^* integral for welded CT samples is proposed.

2. Numerical Analysis

2.1. Geometry and Loading

A two-dimensional (2D) high crack-tip constrained plain-strain CT specimen of welded joints was considered. This specimen consists of three materials, as described in Figure 1. The general geometric dimensions of the CT specimen are also shown in Figure 1. The width of specimen was selected to be $W = 25$ mm. The crack is located in the center of HAZ. For the purpose of the investigation, the relative crack length a/W was varied from 0.5 to 0.9 with a step of 0.1. Additionally, the relative width of the HAZ W/h was varied from 6 to 12 with step 2. All considered specimens were initially loaded with the same stress intensity factor $K = 460$ Nmm$^{-3/2}$.

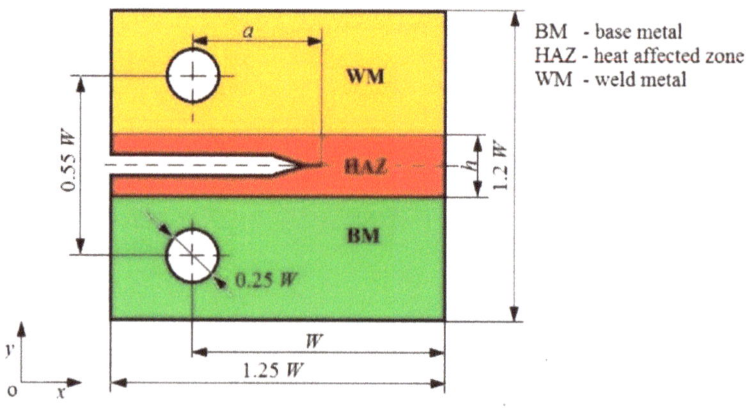

Figure 1. CT specimen of welded joint.

2.2. Material Properties

The elastic-creep material model was used in FE analysis. Generally, the power law relationship is appropriate for modeling the stress–strain response under steady-state creep. Elastic-creep constitutive relation is as follows:

$$\dot{\varepsilon} = \frac{\dot{\sigma}}{E} + A\sigma^n ,$$

(1)

where $\dot{\varepsilon}$. is the uniaxial strain rate, $\dot{\sigma}$. is the uniaxial stress rate, and E is Young's modulus. A and n are the creep constant and exponent, respectively. Young's modules and Poisson's ratio for HAZ and WM are assumed to be the same as those of BM. These values are as follows: E = 175 GPa and v = 0.3. The power law for the three materials (BM, WM, and HAZ) is as follows:

$$\dot{\varepsilon}_{BM} = A_{BM}\sigma^{n_{BM}}, \quad \dot{\varepsilon}_{WM} = A_{WM}\sigma^{n_{WM}}, \quad \dot{\varepsilon}_{HAZ} = A_{HAZ}\sigma^{n_{HAZ}} ,$$

(2)

The selected power law parameters for BM are: $A_{BM} = 1 \times 10^{-20}$ $(\text{MPa})^{-n}\text{h}^{-1}$ and $n_{BM} = 7$. To investigate the influence of material constraints on the C^* integral in welded joints, different configurations of material mismatch in creep strain rate of BM, WM and HAZ were designed. The material creep properties for HAZ and WM were chosen so that the creep strain rate was lower or higher than that of BM. This was achieved by varying the constant A. The creep exponent n for BM, WM and HAZ was identical ($n_{BM} = n_{WM} = n_{HAZ} = 7$). The effect of the mismatch in the creep properties for WM and HAZ relative to BM is expressed by the mismatch factor. These mismatch factors are defined as follows [18]:

$$MF_{WM} = \frac{A_{WM}}{A_{BM}}, \quad MF_{HAZ} = \frac{A_{HAZ}}{A_{BM}}$$

(3)

The selected values of MF_{WM} and MF_{HAZ} factors are 10, 1 and 0.1, respectively. Thus, in order to investigate the effect of material constraints on the C^* integral, nine possible combinations of material mismatch were considered, as shown in Table 1.

Table 1. Considered material mismatches.

MF_{WM}	MF_{HAZ}
10	0.1
10	1
10	10
1	0.1
1	1
1	10
0.1	0.1
0.1	1
0.1	10

2.3. Finite Element Analysis

Extensive elastic-creep FE analyses for welded CT specimens were performed in this study. The general purpose FE program ABAQUS (2016, Dassault Systemes Simulia Corp., Johnston, RI, USA) [20] was used to calculate the C^* integral. A seam crack is included in a 2D CT specimen model. The selected crack front is equivalent to the crack tip. The direction of crack propagation at the crack tip is defined using the virtual crack extension direction. A small geometry change continuum FE model was applied. In order to avoid problems associated with incompressibility, 8 node reduced integration elements for 2D plain-strain problems (CPE8R) were used [3]. The elements of innermost ring at the crack tip are degenerated into triangles. The three nodes along one side of the eight-node element are defined so that they share the same geometrical place. Each of the three collapsed nodes can be displaced independently [21]. Figure 2 shows a typical FE mesh for the CT specimen $a/W = 0.5$ and $W/h = 6$. The crack-tip zone is meshed finely. The mesh size at crack-tip region is selected so as to eliminate mesh-sensitivity in determining the stress fields and C^* integral. Therefore, the element size selected in all further analyses was 0.05 mm.

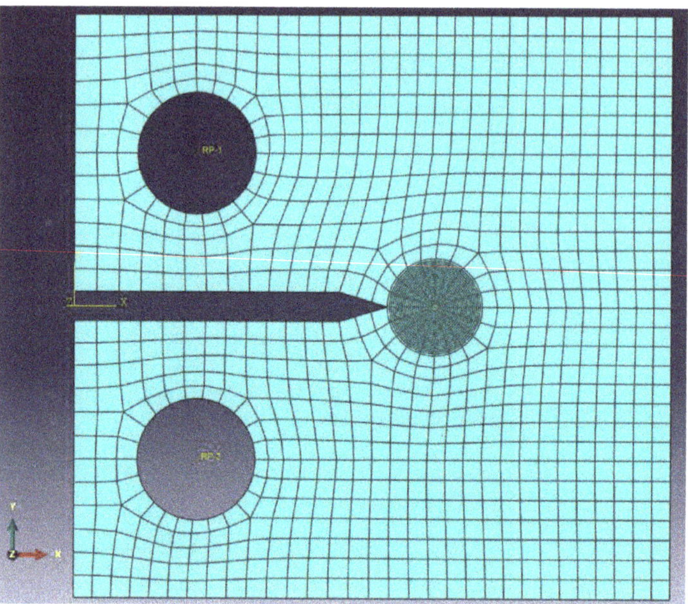

Figure 2. Typical FE mesh for the CT specimen with $a/W = 0.5$.

The load was applied in the center of the upper hole using the reference point RP-1 and the multi-point constraint (MPC) option within ABAQUS. The direction of the load at the reference point was in the y direction. The displacement of the center of the upper hole was constrained in x direction [18]. Additionally, the displacement of the center of the lower hole was constrained in x and y directions [18]. Initially, the model was stress free. The above load was firstly applied instantaneously to the FE model using an elastic calculation at time $t = 0$ [3]. Then, the load was kept constant and a subsequent time-dependent creep calculation was performed. In order to achieve better numerical efficiency for time-dependent creep calculation, a combined implicit and explicit method within ABAQUS was selected. Keeping the load constant, subsequent creep deformation causes stress relaxation at the crack-tip region until a steady-state stress distribution is achieved. The steady-state stress is characterized by path-independent integral C^*. Five different integral paths around the crack tip were considered and the average value of calculated C^*-integrals was taken as final result [21].

To gain confidence in the present FE analysis, elastic and elastic-creep analyses of a homogeneous CT specimen were performed. Figure 3 shows the distribution of von Mises stresses obtained by elastic-creep analysis (deformation scale factor is 10). Table 2 compares the values of fracture mechanics factors K_I and C^* obtained by FE analyses with analytical solutions for CT specimen [22,23]. It is evident that the FE results provide confidence in the FE analyses for elastic and steady-state creep conditions.

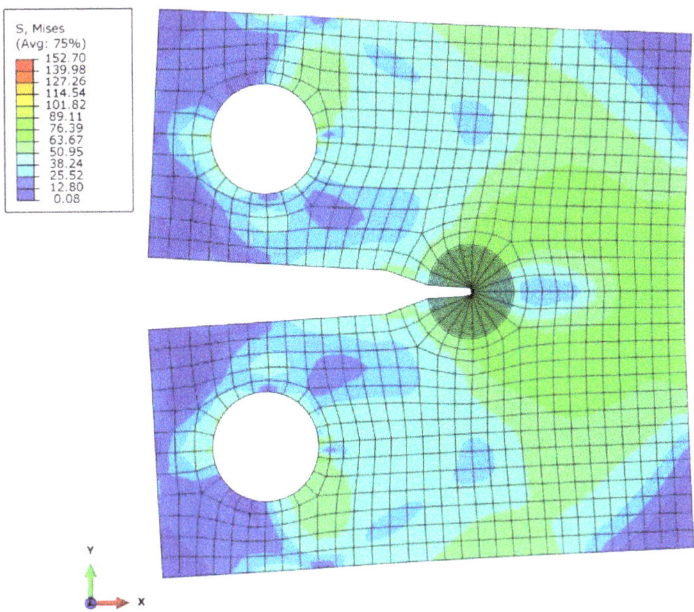

Figure 3. Distribution of von Mises stresses obtained by elastic-creep analysis.

Table 2. Ratios of the FE K_I and C^* with analytical solutions.

$K_I(\text{FE})/K_I$		1.005
$C^*(\text{FE})/C^*$	$n = 5$	0.978
	$n = 10$	0.982

3. Results and Discussion

C^* is estimated experimentally from measurements of creep load line displacement rate according from tests carried out on CT specimens based on the recommendations of ASTM E1457-15. The value of C^* integral is determined using the following Equation [3]:

$$C^* = \eta \frac{n}{n+1} \frac{F\dot{V}_c}{B(W-a)},$$ (4)

where F denotes the applied constant load, B is the specimen thickness, $\dot{V}_c$ is the creep load line displacement rate, and η is an experimental calibration factor. The geometrical factor η depends only on the specimen geometry and the crack length a. For the CT specimen, the geometric factor η is determined according to the following expression:

$$\eta = 2 + 0.522\left(1 - \frac{a}{W}\right)$$ (5)

It is worth noting that Equation (4) is developed for homogeneous materials. This means that this equation is not applicable to heterogeneous materials such as welded joints. A directly measured $\dot{V}_c$ value from the welded CT specimen reflects a partial influence of material and geometry constraints. It is therefore necessary to extend the expression for estimating the C^* integral (Equation (4)) that will take these constraints into account. It can be assumed that the modified equation for estimating the C^* integral for a welded CT specimen has the following form:

$$C^* = \eta \frac{n}{n+1} \frac{F\dot{V}_c}{B(W-a)} \varphi \,, \tag{6}$$

where φ denotes the calibration factor dependent on material and geometry constraints. It can be written as follows:

$$\varphi = \varphi(MF_{WM}, MF_{HAZ}, W/h, a/W) \,, \tag{7}$$

This functional dependence will be found based on the results of extensive elastic-creep FE analyses performed. Factor φ values were calculated as the ratio of the C^* integral of the heterogeneous (welded) CT specimen and the C^* integral for the homogeneous CT specimen:

$$\varphi = \frac{(C^*)_{HET}}{(C^*)_{HOM}} \,, \tag{8}$$

Figure 4a–c show the influence of MF_{HAZ}, MF_{WM} and W/h on the factor φ for the ratio $a/W = 0.5$. It can be seen that MF_{HAZ} has a strong influence on the factor φ for the considered values of MF_{WM} and W/h. Values of φ are highest for creep-soft HAZ ($MF_{HAZ} = 0.1$). It can be seen that these values are significantly higher than one, meaning that the C^* integral is lower than the C^* integral for a homogeneous CT specimen. A larger C^* integral causes a higher rate of CCG. For a given value of MF_{HAZ}, the factor φ is higher if MF_{WM} is lower. For creep-hard HAZ ($MF_{HAZ} = 10$), the values of φ are mostly lower than one, meaning that the C^* integral is less than the C^* integral for a homogeneous CT specimen. The influence of W/h on the φ factor is significant for creep-soft HAZ. The factor φ is higher for the larger HAZ width h (lower W/h ratio).

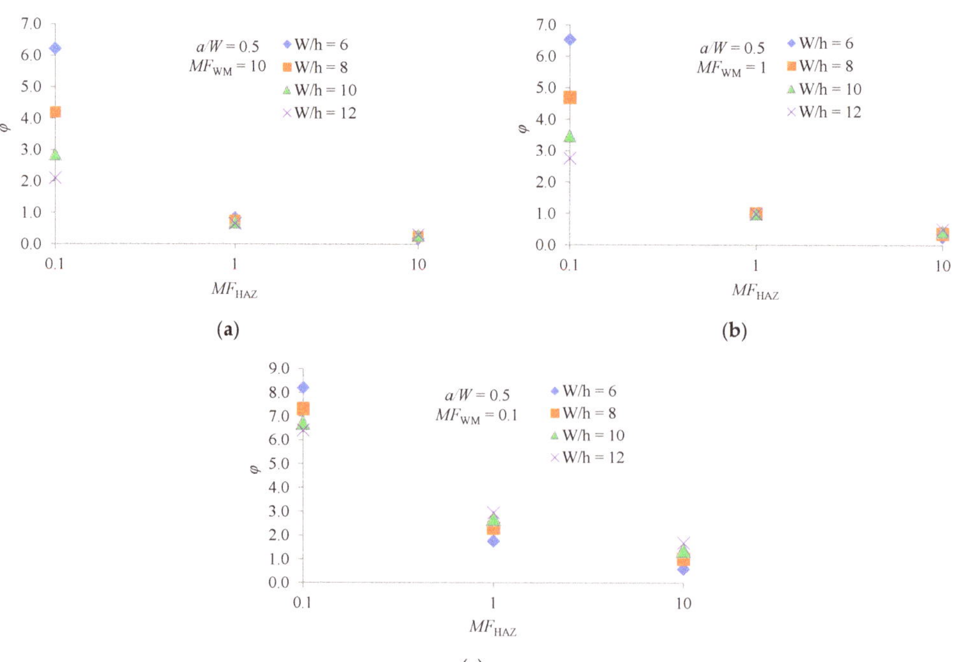

Figure 4. Dependence φ factor on MF_{HAZ}, MF_{WM} and W/h for $a/W = 0.5$: (**a**) $MF_{WM} = 10$; (**b**) $MF_{WM} = 1$; (**c**) $MF_{WM} = 0.1$.

A similar analysis can be performed for the other a/W values considered. However, for easier analysis, Figure 5a–c show the dependence φ factor on the a/W ratio for the considered values of MF_{HAZ} and MF_{WM} at the ratio $W/h = 6$.

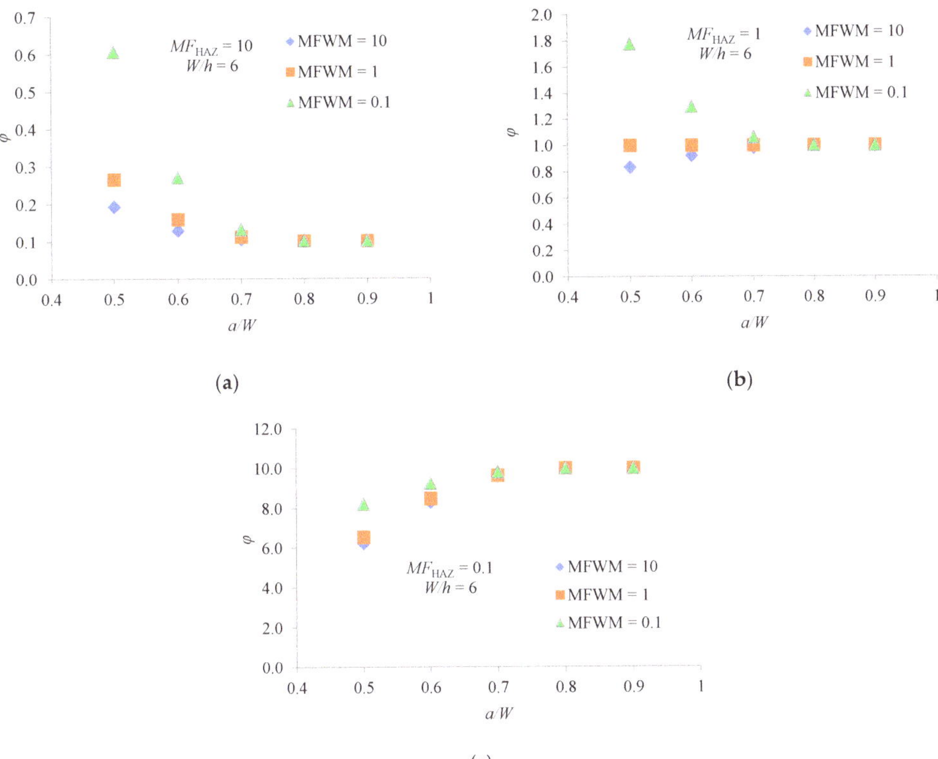

Figure 5. Dependence φ factor on MF_{HAZ}, MF_{WM} and a/W for $W/h = 6$: (**a**) $MF_{HAZ} = 10$; (**b**) $MF_{HAZ} = 1$; (**c**) $MF_{HAZ} = 0.1$.

It can be seen that effects of MF_{HAZ} on φ are quite complex. For creep-hard HAZ ($MF_{HAZ} = 10$), the values of φ are lower than one for all considered values of a/W. The value of φ becomes lower and asymptotically approaches a constant value as a/W increase. The value of MF_{WM} has almost no effect on φ when $a/W \geq 0.7$.

For creep match HAZ ($MF_{HAZ} = 1$), the values of φ in the range $0.5 \leq a/W \leq 0.7$ are higher than one if WM is creep-soft material ($MF_{WM} = 0.1$). On the other side, for creep-hard WM ($MF_{WM} = 10$) and range $0.5 \leq a/W \leq 0.7$ the values of φ are lower than one. In both cases, for values of $a/W > 0.7$, the value of φ asymptotically approaches one. For creep match WM the value of φ is one for all considered values of a/W. This is expected because it is in fact a homogeneous material.

For creep-soft HAZ ($MF_{HAZ} = 0.1$), the values of φ are significantly higher than one for all considered values of a/W. The value of φ becomes higher and asymptotically approaches a constant value as a/W increase. The value of MF_{WM} has almost no effect on φ when $a/W \geq 0.7$.

A similar analysis can be performed for the other W/h values considered. Figure 6a–c show the dependence φ factor on the a/W ratio for the considered values of MF_{HAZ} and MF_{WM} at the ratio $W/h = 8$. Thus, for $a/W = 0.5$ and $MF_{HAZ} = 10$, the corresponding values of φ for $W/h = 8$ are higher than those for $W/h = 6$. Likewise, for $a/W = 0.5$ and $MF_{HAZ} = 0.1$, the corresponding values of φ for $W/h = 8$ are lower than those for $W/h = 6$. In both cases, the value of φ asymptotically approaches a constant value as a/W increase. For $a/W = 0.5$ and $MF_{\text{HAZ}} = 1$ and if WM is creep-soft material ($MF_{\text{WM}} = 0.1$), the value of φ for $W/h = 8$ is higher than this for $W/h = 6$. For $a/W = 0.5$ and $MF_{\text{HAZ}} = 1$ and if WM is creep-hard material ($MF_{\text{WM}} = 10$), the value of φ for $W/h = 8$ is lower than this for $W/h = 6$. In the case where WM and HAZ are creep match materials, the value of φ is one for all considered a/W ratios.

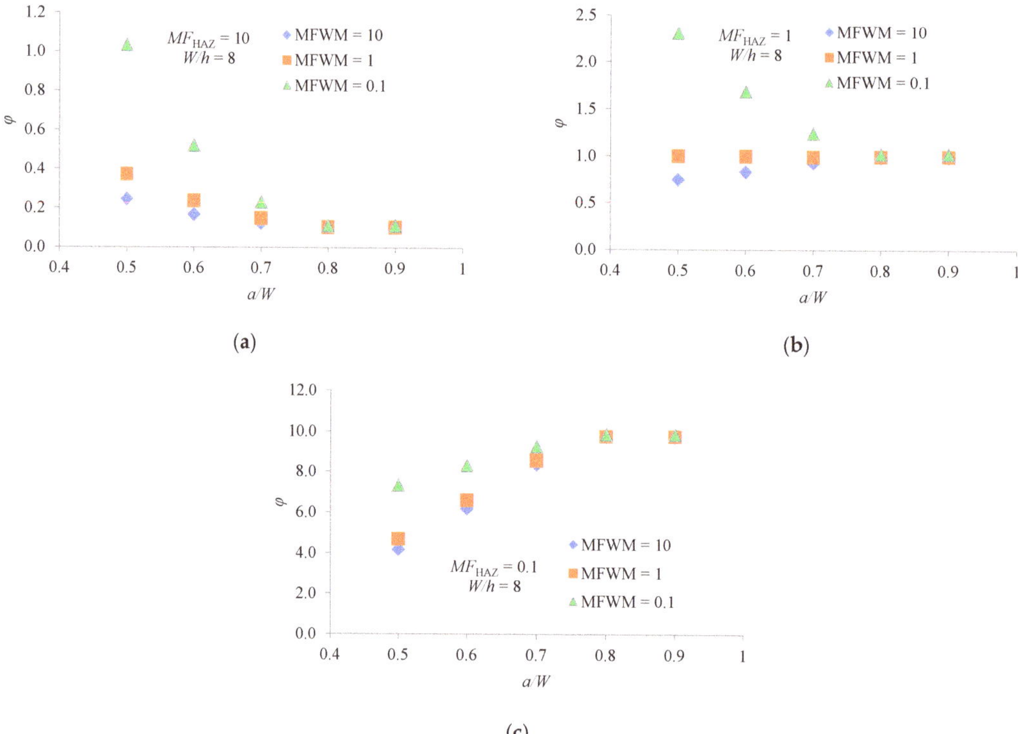

Figure 6. Dependence φ factor on MF_{HAZ}, MF_{WM} and a/W for $W/h = 8$: (**a**) $MF_{\text{HAZ}} = 10$; (**b**) $MF_{\text{HAZ}} = 1$; (**c**) $MF_{\text{HAZ}} = 0.1$.

Considering the dependence of the factor φ on MF_{HAZ}, MF_{WM} and a/W for all W/h ratios, it can be generally concluded that there is a strong and complex influence of material and geometry constraints on the CCG rate. If the HAZ creep is a soft material, which is often the practice [18], the factor φ will have values higher than one. This means that the C^* integral will be higher than the C^* integral for homogenic material, so the CCG rate will also be higher. It is clear that a higher CCG rate means a shorter lifetime of the welded structure.

There is a unique relationship between the C^* integral and the crack-tip stress and strain rate fields as follows [24]:

$$\sigma_{ij} = \left(\frac{C^*}{I_n A r} \right)^{\frac{1}{1+n}} \hat{\sigma}_{ij}(\theta, n)\,, \tag{9}$$

$$\dot{\varepsilon}_{ij} = \left(\frac{C^*}{I_n A r} \right)^{\frac{n}{1+n}} \hat{\varepsilon}_{ij}(\theta, n)\,, \tag{10}$$

where r is distance from the crack tip, I_n is quantity that depends on exponent n and the stress condition, $\hat{\sigma}_{ij}$. and $\hat{\varepsilon}_{ij}$ are angular functions. Analyzing the diagrams in Figures 4–6, it can be clearly concluded that the values of φ change with different combination of MF_{HAZ}, MF_{WM}, W/h and a/W. Different combinations of these parameters change the crack-tip constraint effect, and thus the magnitude of the crack-tip stress and strain rate. A higher magnitude of crack-tip stress and strain rate means a higher value of the C^* integral, and thus a higher value of φ.

Using the results of the analyses performed and applying the software package TuringBot [25], an equation was found which can satisfactorily estimate φ values for the considered values of MF_{HAZ}, MF_{WM}, a/W and W/h. This is the following expression:

$$\varphi = \frac{a/W}{MF_{\mathrm{HAZ}}} + \frac{2}{MF_{\mathrm{HAZ}} + MF_{\mathrm{WM}}(W/h)}, \tag{11}$$

The R-squared for this model is 0.96 and the RMS error is 0.74. This confirms the high goodness-of-fit of the selected model.

4. Conclusions

Based on systematic FE analyses, a modified equation was proposed to estimate the C^* integral for the welded CT specimen with a crack located in the center of the HAZ. Compared to a homogeneous CT specimen, creep-soft HAZ gives higher values of the C^* integral while creep-hard HAZ reduces the value of the C^* integral. This means that the existing equation for the C^* integral in ASTM E 1457 may underestimate or overestimate the actual value of the C^* integral for the welded CT sample. The influence of W/h on the C^* integral is significant for creep-soft HAZ. The C^* integral is higher for the larger HAZ width h (lower W/h ratio). Finally, an expression for estimating the C^* integral for the mismatched welded CT specimen is proposed.

Author Contributions: Conceptualization, M.K. and P.K.; methodology, M.K. and P.K.; validation, M.K., D.T. and P.K.; formal analysis, M.K. and D.T.; investigation, M.K. and D.T.; data curation, M.K. and D.T.; writing—original draft preparation, M.K.; writing—review and editing, M.K., P.K. and D.K.; visualization, M.K.; supervision, D.K. All authors have read and agreed to the published version of the manuscript.

Funding: This research received no external funding.

Institutional Review Board Statement: Not applicable.

Informed Consent Statement: Not applicable.

Conflicts of Interest: The authors declare no conflict of interest.

References

1. Dake, Y.; Sridrah, I.; Zhongmin, Z.; Kumar, S.B. Fracture capacity of girth welded pipelines with 3D surface cracks subjected to biaxial loading conditions. *Int. J. Press. Vessels Pip.* **2012**, *92*, 115–126. [CrossRef]
2. Dake, Y.; Zhong, M.X.; Sridrah, I.; Kumar, S.B. Fracture analysis of girth welded pipelines with 3D embedded cracks subjected to biaxial loading conditions. *Eng. Fract. Mech.* **2012**, *96*, 570–587.
3. Xuan, F.-Z.; Tu, S.-T.; Wang, Z. A modification of ASTM E 1457 C^* estimation equation for compact tension specimen with a mismatched cross-weld. *Eng. Fract. Mech.* **2005**, *72*, 2602–2614. [CrossRef]
4. Abe, F.; Tabuchi, M.; Tsukamoto, S.; Shirane, T. Microstructure evolution in HAZ and suppression of Type IV fracture in advanced ferritic power plant steels. *Int. J. Press. Vessel. Pip.* **2010**, *87*, 598–604. [CrossRef]
5. Dogan, B. *Proceedings of CREEP8, Eighth International Conference on Creep and Fatigue at Elevated Temperatures (Joint with PVP07)*; University of Limerick: Limerick, Ireland, 2007; pp. 299–305.
6. Tabuchi, M.; Hongo, H.; Watanabe, T., Jr.; Yokobori, A.T. Creep crack growth analysis of welded joints for high Cr heat resisting steel. *ASTM Spec. Tech. Publ.* **2007**, *1480*, 93–101.
7. Yatomi, M.; Fuji, A.; Tabuchi, M.; Hasegawa, Y.; Kobayashi KI, I.; Yokobori, T.; Yokobori, T. Creep crack growth of P92 welds. In Proceedings of the ASME 2008 Pressure Vessels and Piping Division Conference, Chicago, IL, USA, 27–31 July 2008; pp. 1207–1214.
8. Hyde, T.; Saber, M.; Sun, W. Testing and modelling of creep crack growth in compact tension specimens from a P91 weld at 650 °C. *Eng. Fract. Mech.* **2010**, *77*, 2946–2957. [CrossRef]

9. Lee, K.-H.; Kim, Y.-J.; Yoon, K.-B.; Nikbin, K.; Dean, D. Quantification of stress redistribution due to mismatch in creep properties in welded branch pipes. *Fatigue Fract. Eng. Mater. Struct.* **2010**, *33*, 238–251. [CrossRef]

10. Tu, S.-T.; Segle, P.; Gong, J.-M. Creep damage and fracture of weldments at high temperature. *Int. J. Press. Vessel. Pip.* **2003**, *81*, 199–209. [CrossRef]

11. Tu, S.-T.; Yoon, K.-B. The influence of material mismatch on the evaluation of time-dependent fracture mechanics parameters. *Eng. Fract. Mech.* **1999**, *64*, 765–780. [CrossRef]

12. Xuan, F.-Z.; Wang, Z.-F.; Tu, S.-T. Creep finite element simulation of multilayered system with interfacial cracks. *Mater. Des.* **2009**, *30*, 563–569. [CrossRef]

13. Chen, G.; Wang, G.; Xuan, F.; Tu, S. Mismatch effect in creep properties on creep crack growth behavior in welded joints. *Mater. Des.* **2014**, *63*, 600–608. [CrossRef]

14. Chen, G.; Wang, G.; Zhang, J.; Xuan, F.; Tu, S. Effects of initial crack positions and load levels on creep failure behavior in P92 steel welded joint. *Eng. Fail. Anal.* **2015**, *47*, 56–66. [CrossRef]

15. Chen, G.; Wang, G.Z.; Xuan, F.Z.; Tu, S.T. Effects of HAZ widths on creep crack growth properties of welded joints. *Weld. World* **2015**, *59*, 851–860. [CrossRef]

16. Zhao, L.; Jing, H.; Xiu, J.; Han, Y.; Xu, L. Experimental investigation of specimen size effect on creep crack growth behavior in P92 steel welded joint. *Mater. Des.* **2014**, *57*, 736–743. [CrossRef]

17. Mehmanparast, A.; Maleki, S.; Yatomi, M.; Nikbin, K.M. Specimen Geometry and Size Effects on the Creep Crack Growth Behaviour of P91 Weldments. In Proceedings of the ASME 2013 Pressure Vessels and Piping, Division Conference, Paris, France, 14–18 July 2013.

18. Li, Y.; Wang, G.Z.; Xuan, F.Z.; Tu, S.T. Geometry and Material Constraint Effects on Creep Crack Growth Behavior in Welded Joints. *High Temp. Mater. Process.* **2016**, *36*, 155–162. [CrossRef]

19. ASTM Standard, E1457-15. *Standard Test Method for Measurement of Creep Crack Growth Times in Metals*; ASTM International: West Conshohocken, PA, USA, 2015.

20. ABAQUS /CAE 2016 User's Manual. Inc. and Dassault Systems. Available online: http://130.149.89.49:2080/v2016/index.html (accessed on 15 June 2021).

21. Katinić, M.; Konjatić, P.; Kozak, D.; Turk, D. Numerical analysis of the effect of residual stress on transient creep in SENB specimen. *Int. J. Press. Vessels Pip.* **2020**, *188*, 104222. [CrossRef]

22. Tada, H.; Paris, P.; Irwin, G. *The Stress Analysis of Cracks Handbook*; Paris Productions Inc.: St. Louis, MO, USA, 1985.

23. Kumar, V.; German, D.; Shih, C.F. *An Engineering Approach for Elastic-Plastic Fracture Analysis*; EPRI Report NP-1931; EPRI: Washington, DC, USA, 1981.

24. Goldman, N.L.; Hutchinson, J.W. Fully Plastic Crack Problems: The Center Cracked Strip Under Plane Strain. *Int. J. Solids Struct.* **1975**, *11*, 575–591. [CrossRef]

25. TuringBot Software, Sao Paulo, Brazil. 2021. Available online: https://turingbotsoftware.com/ (accessed on 12 August 2021).

Article

Effects of Fixture Configurations and Weld Strength Mismatch on *J*-Integral Calculation Procedure for SE(B) Specimens

Primož Štefane [1], Stijn Hertelé [2], Sameera Naib [2], Wim De Waele [2] and Nenad Gubeljak [1,*]

[1] Faculty of Mechanical Engineering, University of Maribor, 2000 Maribor, Slovenia; primoz.stefane2@um.si
[2] Soete Laboratory, EMSME Department, Faculty of Engineering and Architecture, Ghent University, 9000 Ghent, Belgium; stijn.hertele@ugent.be (S.H.); sameera.naib@ugent.be (S.N.); Wim.DeWaele@UGent.be (W.D.W.)
*   Correspondence: nenad.gubeljak@um.si; Tel.: +386-2-220-7661

Abstract: This work presents the development of a *J*-integral estimation procedure for deep and shallow cracked bend specimens based upon plastic η_{pl} factors for a butt weld made in an S690 QL high strength low alloyed steel. Experimental procedures include the characterization of average material properties by tensile testing and evaluation of base and weld metal resistance to stable tearing by fracture testing of square SE(B) specimens containing a weld centerline notch. *J*-integral has been estimated from plastic work using a single specimen approach and the normalization data reduction technique. A comprehensive parametric finite element study has been conducted to calibrate plastic factor η_{pl} and geometry factor λ for various fixture and weld configurations, while a corresponding plastic factor γ_{pl} was computed on the basis of the former two. The modified η_{pl} and γ_{pl} factors were then incorporated in the *J* computation procedure given by the ASTM E1820 standard, for evaluation of the plastic component of *J* and its corresponding correction due to crack growth, respectively. Two kinds of *J-R* curves were computed on the basis of modified and standard η_{pl} and γ_{pl} factors, where the latter are given by ASTM E1820. A comparison of produced *J-R* curves for the base material revealed that variations in specimen fixtures can lead to ≈10% overestimation of computed fracture toughness J_{Ic}. Furthermore, a comparison of *J-R* curves for overmatched single-material idealized welds revealed that the application of standard η_{pl} and γ_{pl} factors can lead to the overestimation of computed fracture toughness J_{Ic} by more than 10%. Similar observations are made for undermatched single material idealized welds, where fracture toughness J_{Ic} is overestimated by ≈5%.

Keywords: metal weld; strength mismatch; fracture; plastic correction factors; fixture rollers; *J-R* resistance curve

Citation: Štefane, P.; Hertelé, S.; Naib, S.; De Waele, W.; Gubeljak, N. Effects of Fixture Configurations and Weld Strength Mismatch on *J*-Integral Calculation Procedure for SE(B) Specimens. *Materials* **2022**, *15*, 962. https://doi.org/10.3390/ma15030962

Academic Editor: Tomasz Trzepieciński

Received: 30 November 2021
Accepted: 19 January 2022
Published: 26 January 2022

Publisher's Note: MDPI stays neutral with regard to jurisdictional claims in published maps and institutional affiliations.

1. Introduction

The fracture resistance (in the form of a *J-R* curve) of the weld metal is an essential input for structural integrity assessments of load bearing welded components according to various fitness for service (FFS) assessment methods (e.g., BS7910 [1], R6 [2] and FIT-NET [3,4]). Experimental determination of the *J-R* curve for welded joints is based on the testing of small, laboratory fracture specimens according to standardized procedures, specified by ISO 15,653 [5] and BS 7448-2 [6]. The former is based on the *J*-evaluation method for homogeneous metallic materials, included in ASTM E1820 [7], which has been extended to weldments with yield strength mismatch ratio *M* in the range $0.5 \leq M \leq 1.25$. Here, *M* is defined as:

$$M = \frac{\sigma_{yWM}}{\sigma_{yBM}}, \tag{1}$$

where σ_{yWM} and σ_{yBM} are all-weld metal and base metal yield strength, obtained by tensile testing. In case *J*-integral is evaluated by a single specimen approach, incremental equations are used where elastic and plastic contributions to the strain energy of the cracked specimen are considered according to standard ASTM E1820 as follows:

$$J_{(i)} = J_{el(i)} + J_{pl(i)}, \tag{2}$$

where $J_{(i)}$ is the total J-integral while $J_{el(i)}$ and $J_{pl(i)}$ are elastic and plastic components of the total J-integral respectively. The present study is focused on the determination of J by testing of single edge notched bend (SE(B)) specimens. Therefore, J computation equations for SE(B) specimens will be discussed throughout this paper. The elastic component $J_{el(i)}$ in Equation (2) for plane strain is given by:

$$J_{el(i)} = \frac{K_{I(i)}^2}{E'} = \frac{K_{I(i)}^2\left(1 - v^2\right)}{E}, \tag{3}$$

where, $E' = E/\left(1 - v^2\right)$, E is the elastic modulus, v is Poisson's ratio and K_I is the stress intensity factor (SIF) for mode I crack opening, that is:

$$K_{I(i)} = \frac{P_{(i)}S}{(BB_N)^{1/2}W^{3/2}} \cdot f\left(a_{(i)}/W\right), \tag{4}$$

where $P_{(i)}$ and $a_{(i)}$ are the applied load and crack size in the considered time increment, W, B and B_N denote width, thickness and net thickness (smaller than B if the specimen is side-grooved, otherwise $B = B_N$) of the SE(B) specimen and S denotes the span length between the support rollers. The function $f(a_i/W)$ is a nondimensional factor that depends on crack size $a_{(i)}$ normalized by specimen width W, and is given by:

$$f\left(a_{(i)}/W\right) = \frac{3\left(\frac{a_i}{W}\right)^{1/2}\left[1.99 - \frac{a_i}{W}\left(1 - \frac{a_i}{W}\right)\left(2.15 - 3.93\left(\frac{a_i}{W}\right) + 2.7\left(\frac{a_i}{W}\right)^2\right)\right]}{2\left(1 + 2\frac{a_i}{W}\right)\left(1 - \frac{a_i}{W}\right)^{3/2}}. \tag{5}$$

The plastic component of the J-integral, $J_{pl(i)}$, is evaluated by correlation with the area under the load-plastic displacement curve. An incremental equation for computation of $J_{pl(i)}$ was originally proposed by Zhu et al. [8] and is included in ASTM E1820 in the following form

$$J_{pl(i)} = \left[J_{pl(i-1)} + \left(\frac{\eta_{pl(i-1)}}{b_{(i-1)}}\right)\left(\frac{A_{pl(i)} - A_{pl(i-1)}}{B_N}\right)\right] \cdot \left[1 - \gamma_{pl(i-1)}\left(\frac{a_{(i)} - a_{(i-1)}}{b_{(i-1)}}\right)\right]. \tag{6}$$

Here, b is remaining ligament ($b = W - a$) and A_{pl} is the area under the load-plastic displacement curve defined by the incremental trapezoidal integration rule:

$$A_{pl(i)} = A_{pl(i-1)} + \left[P_{(i)} + P_{(i-1)}\right] \cdot \left[V_{pl(i)} - V_{pl(i-1)}\right]/2. \tag{7}$$

Here, V_{pl} denotes the plastic component of the measured displacement, which is the crack mouth opening displacement (CMOD) in this paper. Equations (8) and (9) determine η_{pl} and γ_{pl} factors, required to calculate the J-integral from the load-CMOD record using Equation (6). The former relates to the area under the load- plastic CMOD curve while the latter relates to the incremental plastic work for crack growth. Both factors are functions that depend on normalized crack size for SE(B) specimens.

$$\eta_{pl(i-1)} = 3.667 - 2.199\left(\frac{a_{(i-1)}}{W}\right) + 0.437\left(\frac{a_{(i-1)}}{W}\right)^2 \tag{8}$$

$$\gamma_{pl(i-1)} = 0.131 + 2.131\left(\frac{a_{(i-1)}}{W}\right) - 1.465\left(\frac{a_{(i-1)}}{W}\right)^2. \tag{9}$$

Notably, in fracture toughness testing, CMOD is normally preferred over load line displacement (LLD) because a dedicated CMOD gage is simpler to use in an experiment than a complex LLD gage [9]. Furthermore, a study of Kirk and Dodds [10] provided results which showed that the LLD based J-integral estimation procedure gives accurate results for $a/W > 0.3$, but inaccurate results for $a/W < 0.3$ due to a high sensitivity of the η_{pl} factor to the strain hardening exponent of the material for SE(B) specimens with shallow cracks. In contrast, the CMOD based η_{pl} factor is insensitive to the strain hardening for $a/W > 0.05$ for SE(B) specimens. Fixtures as devices for accurate locating and reliable support during bending testing should be sized according to the ASTM E1820 standard. According to the mentioned standard, rollers should be free in order to keep a constant loading arm. Fixed rollers can have an influence on results, as will be discussed in this paper.

In the past, studies to improve the fracture testing method, based on SE(B) specimens with $W/B = 2$ configuration, have introduced improved η_{pl} factors [10–12]. The solutions included in ASTM E1820, in the form of expressions (8) and (9) for computation of η_{pl} and γ_{pl} respectively for SE(B) specimens, were proposed by Zhu et al. [8] based on finite element results published by Donato and Ruggieri [12]. Both expressions are assumed to be accurate for a range of crack sizes $0.25 \leq a/W \leq 0.7$ [12], while the range of validity is reduced to $0.45 \leq a/W \leq 0.7$ in ASTM E1820. As discussed, ISO 15653 assumes that expression (8) is valid for M values in the range $0.5 \leq M \leq 1.25$. However, M values of weld joints used for various applications often exceed the limit of 1.25. It is necessary to adopt appropriate η_{pl} and γ_{pl} equations in terms of strength mismatch M and normalized crack length a/W, in order to accurately evaluate J-R curves for such joints.

Several researchers provided η_{pl} and γ_{pl} solutions for the evaluation of J in fracture testing of strength mismatched weld joints. Kim et al. [13] performed detailed finite element (FE) analyses to obtain η_{pl} of various specimens (including SE(B)) with $a/W = 0.5$ for weld joints with strength mismatch M varying between 0.5 and 2.0. The obtained results demonstrated that values of η_{pl} increase in case of weld strength undermatching ($M < 1$) and decrease in case of weld strength overmatching ($M > 1$), relative to even matching welds ($M = 1$). Furthermore, it was shown that η_{pl} depends on the weld width. Starting from a very wide weld, the reduction of the weld width results in an increase of η_{pl} values for undermatching welds, reaching a maximum value at geometry ratio $(W - a)/H_W = 5$, where H_W is weld width. The opposite was observed for overmatching welds, where reduction of the weld width resulted in an decrease of η_{pl} values, reaching a minimum value at geometry ratio $(W - a)/H_W = 2$. For very narrow welds with $(W - a)/H_W < 2$, the effect of the weld geometry was negligible and computed values of η_{pl} were similar to the one for pure base metal. Eripret and Hornet [14] performed a parametric FE study of SE(B) specimens from weld joints with mismatch levels $M = 0.2$ and $M = 2.0$ for a wide range of crack lengths ($0.1 \leq a/W \leq 0.7$). Results demonstrated that η_{pl} values increase across the entire range of analysed crack lengths by reducing the mismatch factor M. Similar observations were made by Donato et al. [15]. Their study showed that scatter of produced η_{pl} solutions for analysed levels of M is relatively small if the crack is located at the central plane of a narrow weld with $H_W = 5$ mm and relatively high in a wider weld with $H_W = 20$ mm (SE(B) specimens with $B = 25.4$ mm and $W = 2B = 50.8$ mm). While aforementioned studies [13–15] incorporated 2D plane strain conditions in parametric FEM for SE(B) specimens, Mathias et al. [16] performed parametric 3D FE analyses of SE(B) specimens with $W/B = 1$ and $W/B = 2$ for a wide range of crack lengths ($0.1 \leq a/W \leq 0.7$). Although the SE(B) samples were from an X80 steel welded joint with $M = 1.18$ (according to published yield stresses for base and weld material), η_{pl} solutions were developed for homogeneous material with various yield strength levels and hardening properties. Results revealed that the produced η_{pl} solution for SE(B) specimens with $W/B = 2$ is in close agreement with the one obtained by Donato [12], while the η_{pl} solution for SE(B) specimens with $W/B = 1$ was considerably lower (approx. 11% for shallow cracks and 25% for deep cracks).

The above mentioned reported effects of SE(B) sample geometry, weld size and weld strength mismatch on the J evaluation procedure are the main motivation for this research. The work focuses on the development of n_{pl} solutions for SE(B) specimens, extracted from S690 QL steel weld joints with various mismatch levels M. Welding procedures, determination of weld and base material mechanical properties, fracture testing and the calibration of η_{pl} and γ_{pl} functions by parametric FEM are presented in the following paragraphs.

2. Experimental Procedures

2.1. Materials and Welding

Two types of welded plates, as shown in Figures 1 and 2, were fabricated, joining 25 mm thick high strength low alloyed (HSLA) steel S690 QL plates with 500 mm length and 200 mm width, by metal active gas (MAG) welding, with the purpose of extracting specimens for fracture testing and tensile testing. Parent plates during welding were not fixed. For the first type, the weld groove had been machined to a double V configuration with a bevel angle of 60° and root gap of 2 mm; a commonly used weld configuration in practice. For the second type, the weld groove had been machined to a wide V configuration with bevel angle of 20° and weld root gap of 20 mm. In this case, a 10 mm thick backing strip had been attached to both base metal plates to be joined beneath the weld groove in order to fabricate the weld. Such weld configuration meets the requirements of standard ISO 15792-1 [17] for tensile testing of weld consumables and extraction of corresponding tensile test specimens. Weld consumables Mn4Ni2CrMo (with commercial designation MIG 90) and G4Si1 (with commercial designation VAC 65) have been applied in order to fabricate overmatched (OM) welds with $M > 1$ and undermatched (UM) welds with $M < 1$ respectively.

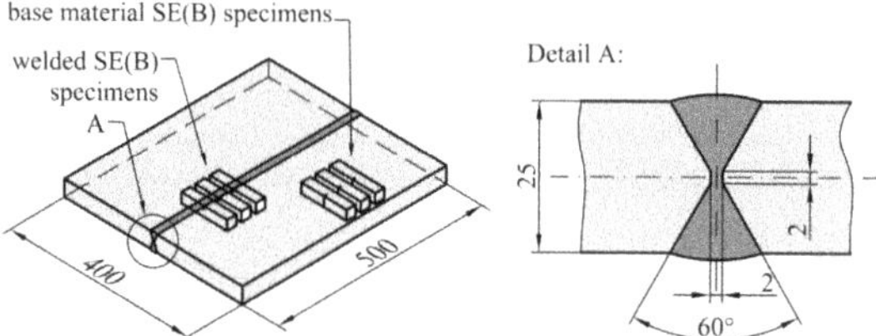

Figure 1. Drawing of the fabricated welded plate for extraction of SE(B) specimens. Weld geometry and layout of the SE(B) specimens is the same for overmatched and undermatched weld.

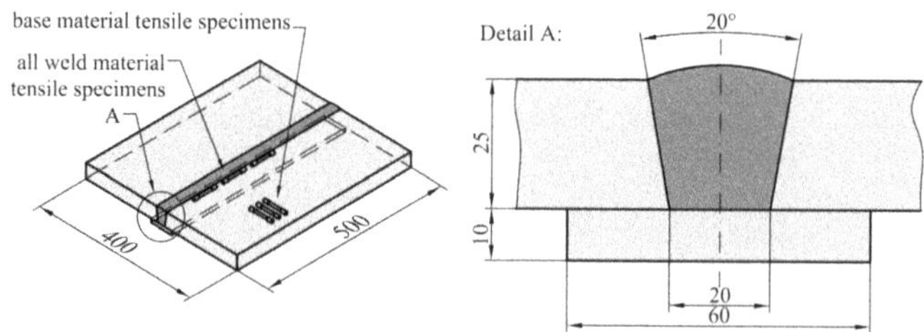

Figure 2. Drawing of the fabricated welded plate for extraction of tensile specimens. Weld geometry and layout of the tensile specimens is the same for overmatched and undermatched weld.

2.2. Tensile Testing

Tensile testing of base and all-weld metals was performed in conformance with ASTM E8/E8M [18], utilizing round bar tensile specimens with neck diameter $D = 6$ mm and gauge length $G = 5D = 30$ mm. Tensile specimens were tested on a multipurpose testing machine INSTRON 1255 by applying displacement-controlled loading with a crosshead displacement rate of 0.2 mm/min at room temperature. Three tensile tests were performed for each material. The obtained average tensile properties are presented in Table 1, where E is the elastic modulus, σ_{YS} and σ_{UTS} are yield strength and ultimate tensile strength respectively and M is the mismatch factor defined by Equation (1). The obtained average engineering stress-strain (S-e) curves are presented in Figure 3.

Table 1. Tensile properties of tested materials.

Material	E (GPa)	σ_{YS} (MPa)	σ_{UTS} (MPa)	M (-)
Base material (S690 QL)	201	683	791	1
OM weld material (Mn4Ni2CrMo)	215	894	950	1.309
UM weld material (G4Si1)	210	532	587	0.779

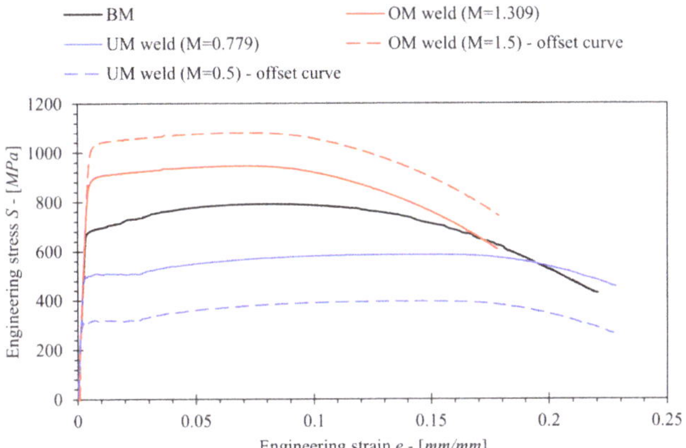

Figure 3. Engineering stress-strain curves of tested materials.

Furthermore, fictitious stress–strain curves were computed by offsetting the S-e curves for OM and UM welded joints along the slope of the linear elastic part in order to achieve mismatch factors $M = 1.5$ and $M = 0.5$ respectively. Experimental and fictitious S-e curves were then implemented in FEM in order to investigate the influence of mismatching on solutions for the J computation factors η_{pl}, λ and γ_{pl}.

2.3. Fracture Toughness Testing

Fracture toughness testing of base material and welds was performed by the single specimen test method according to ASTM E1820 [7]. SE(B) specimens with thickness and width $B = W = 20$ mm have been tested as shown in Figure 4. Three sets of SE(B) specimens were extracted from sample plates with the double V weld configuration for fracture toughness testing of base material, OM weld and UM weld, as shown in Figure 1. Side surfaces of extracted SE(B) specimens containing a weld were first ground and then etched with a 4 % nitric acid alcohol solution in order to determine the position of the weld center line and the weld fusion lines. Finally, notches and aligned knife edges were fabricated at the weld center by wire electrical discharge machining. Weld SE(B) specimens were notched in the direction of the plate thickness (i.e., obtaining surface cracked welded SE(B) specimens). Such notch configuration proved to be less demanding for fatigue precracking, as surface cracked welded SE(B) specimens (with dimensions $B = W$) normally do not exhibit a non-uniform fatigue crack front [19] (maximum relative deviations from computed average fatigue crack length a_0 are reported in Table 2). Therefore, no adapted precracking procedure for modification of residual stresses is required. Sharp cracks were introduced in

SE(B) specimens through fatigue precracking with load ratio $R = 0.1$ and applied maximum SIF to elastic modulus ratio $K_{max}/E \leq 1.1 \times 10^{-4}$ m$^{1/2}$.

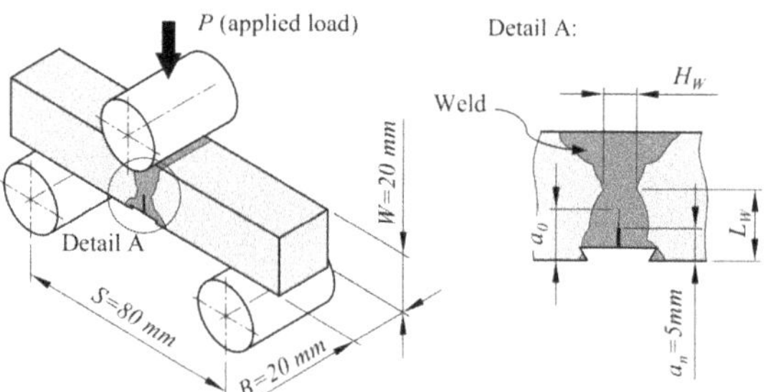

Figure 4. Example of tested weld joint SE(B) specimen with a surface notch at the weld center plane. Missing dimensions L_W, H_W and a_0 are listed in Table 2.

Table 2. List of tested SE(B) fracture specimens with corresponding materials and crack lengths.

Specimen	Material	a_0 [mm]	a_0/W [-]	a_f [mm]	Δa [mm]	L_W [mm]	L_W/W [mm]
SE(B)-02	Base material (S690 QL)	10.645 (7.5%) *	0.533	11.785	1.140	-	-
SE(B)-38	OM weld material (Mn4Ni2CrMo)	8.579 (7.2%) *	0.431	9.888	1.309	7.200	0.360
SE(B)-40	UM weld material (G4Si1)	9.015 (4.2%) *	0.448	9.995	0.980	8.300	0.415

Remarks: *—Maximum relative deviation of 9 measured crack lengths from the average computed crack length (acceptable as specified in ASTM E1820).

The tests were performed on a multipurpose testing machine INSTRON 1255 under crosshead displacement control at displacement rate 1 mm/min and room temperature. A fixture system with fixed support and load rollers with diameter of 25 mm was used. CMOD was measured with a dedicated clip gauge mounted onto the specimen surface. Stable crack extension in SE(B) specimens was quantified by combining post-mortem crack measurements using the nine-point method and the normalization data reduction (NDR) method as specified in ASTM E1820. Although the NDR method is normally used for stable crack extension estimation in homogeneous metallic materials [20], a recent study conducted by Tang et al. [21] showed that it can be applied to welds as well. An average difference of less than 10% between the *J-R* resistance curves provided by the NDR method and unloading compliance method was reported in the listed study. The overview of tested specimens is provided in Table 2. Fracture testing results are presented as load-CMOD curves in Figure 5. Material resistance to fracture in form of *J-R* curves are presented and discussed in detail in Section 6, which includes the effect of various iterations of η_{pl} and γ_{pl} functions on the computed *J*-integral.

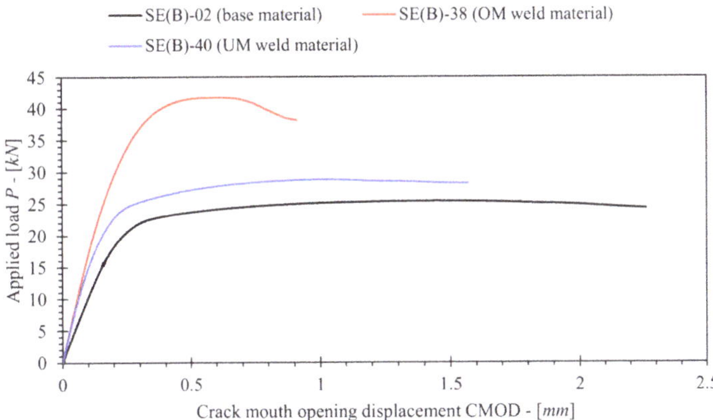

Figure 5. Example of tested weld joint SE(B) specimen with a surface notch at the weld center plane. Missing dimensions L_W, H_W and a_0 are listed in Table 2.

3. Numerical Procedures

3.1. Weld Geometry Simplification Procedure

Current fracture assessment procedures adopt an idealized weld geometry with straight fusion lines to represent more complex weld configurations found in engineering applications. A systematic methodology for simplification of an actual V-groove weld with a centerline crack to an idealized weld has been proposed by Hertelé et al. [22,23]. This methodology has been developed for single edge tension (SE(T)) specimens and is based on the analysis of slip-line patterns. Research conducted by Souza et al. [24] showed that the weld simplification methodology proposed by Hertelé et al. [22] is adequate for V-grooved welds with straight fusion lines for various weld strength mismatch levels. However, the proposed methodology fails to produce accurate results in the presence of high levels of weld strength undermatch as the deformation pattern near the crack tip changes significantly. Considering this, the double-V weld was in the scope of this research simplified to have bi-linear fusion lines rather than a square weld cross section geometry consisting of perfectly straight fusion lines (Figure 6).

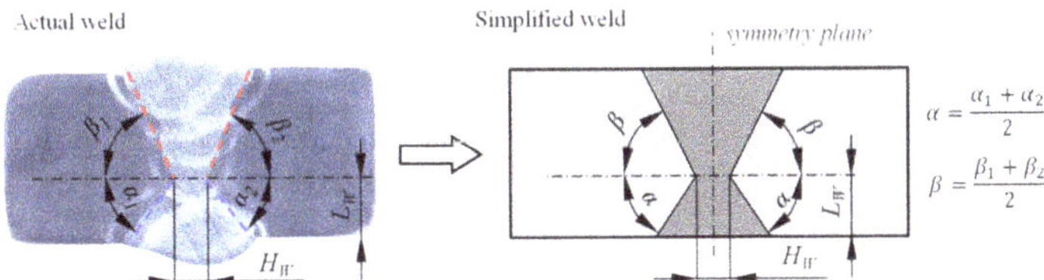

$$\alpha = \frac{\alpha_1 + \alpha_2}{2}$$

$$\beta = \frac{\beta_1 + \beta_2}{2}$$

Figure 6. Approach to derive a simplified weld from an irregularly shaped weld.

The geometry of overmatched and undermatched welds was modelled symmetrically with respect to the central vertical plane of the SE(B) specimen. Simplification of the double-V weld geometry has been done through post-processing of digital macrographs of actual welds using the following procedure. First, four distinctive fusion lines were recognized with respect to the position of the weld root; upper right, upper left, lower right and lower left fusion line. Points were then marked along each of the four fusion lines and coordinates of the marked points were extracted. Next, straight fusion lines were fitted to the extracted coordinates using linear regression. Average slopes of upper and lower fusion lines have been computed in order to create a symmetrical simplified geometry of the weld. The width of the weld root has been measured in order to accurately adjust the minimum width of the idealized weld in FEM. Finally, the side surfaces of actual SE(B) specimens have been etched and the vertical position of the weld root was measured prior to fracture testing. The position of the weld root was later transferred to the idealized weld in FEM. This way, an idealized weld for each parametric FEM series was adjusted to match a corresponding SE(B) specimen that underwent fracture testing. The simplified weld geometry is shown in Figure 6, while corresponding dimensions are presented in Table 3.

Table 3. Computed dimensions of simplified welds.

Weld	α [°]	β [°]	H_W [mm]	L_W [mm]
OM weld material (Mn4Ni2CrMo)	48.3	62.7	5.17	7.2
UM weld material (G4Si1)	42.9	51.7	3.47	8.3

3.2. Numerical Models of Tested Specimens

Detailed nonlinear finite element analyses have been performed using ABAQUS 2018. Plane strain finite element models have been created for a wide range of SE(B) specimens with width W = 20 mm, width to thickness ratio W/B = 1 and span length S = $4W$ = 80 mm. The analysis matrix shown in Table 4 includes seven distinctive FEM series of SE(B) specimens containing base material and idealized overmatched and undermatched welds in combination with three different support and load roller setups.

Table 4. The analysis matrix with FEM series distinctive features.

FEM Series	Material	Support and Load Rollers Diameters and Degrees of Freedom	Modelled Normalized a_0/W Crack Lengths
1a	base material	d_S = 10 mm (free in horizontal plane) d_L = 8 mm (applied displacement in vertical plane)	
1b		d_S = 10 mm (free in horizontal plane) d_L = 25 mm (applied displacement in vertical plane)	0.1, 0.15, 0.2, 0.25, 0.3, 0.4, 0.5, 0.6, 0.7
1c		d_S = 25 mm (fixed) d_L = 25 mm (applied displacement in vertical plane)	
2a	OM weld M = 1.302 (L_0/W = 0.36)	d_S = 25 mm (fixed) d_L = 25 mm (applied displacement in vertical plane)	0.1, 0.15, 0.2, 0.25, 0.3, 0.36, 0.4, 0.5, 0.6, 0.7
2b	OM weld M = 1.5 (L_0/W = 0.36)		
3a	UM weld M = 0.779 (L_0/W = 0.415)	d_S = 25 mm (fixed) d_L = 25 mm (applied displacement in vertical plane)	0.1, 0.15, 0.2, 0.25, 0.3, 0.415, 0.5, 0.6, 0.7
3c	UM weld M = 0.5 (L_0/W = 0.415)		

Figure 7a,b show examples of plane strain FEM models for SE(B) specimens with a_0/W = 0.5 consisting of base metal and containing a welded joint (similar geometry for overmatched and undermatched welds). All other models have similar features. A conventional mesh configuration having a focused ring of finite elements surrounding a stationary crack with a blunted tip. According to SIMULIA documentation [25], contour integral (that is J in this paper) should be accurately evaluated if radius of blunted crack tip is $\rho_0 \approx 10^{-3} r_p$. Here, r_p is size of plastic zone ahead of the crack tip that is determined according to Irwin [26] as:

$$r_p = \frac{1}{2\pi}\left(\frac{K_I}{\sigma_{YS}}\right)^2,$$

(10)

where K_I is SIF that is obtained by post-processing of recorded P-CMOD curves (Figure 5) using 95% secant method as specified in ASTM E399 [27]. Size of plastic zone has been estimated for each tested material using Equation (10). Results presented in Table 5 suggest that average crack tip radius $\approx$1.5 μm could be modelled in all FEM in order to minimize the influence of geometry and mesh on computed results. However, blunt crack tip radius

$\rho_0 = 2.5$ μm was implemented in analyzed FEM as published studies reported that such stationary crack configuration produces sufficiently accurate results [12,15].

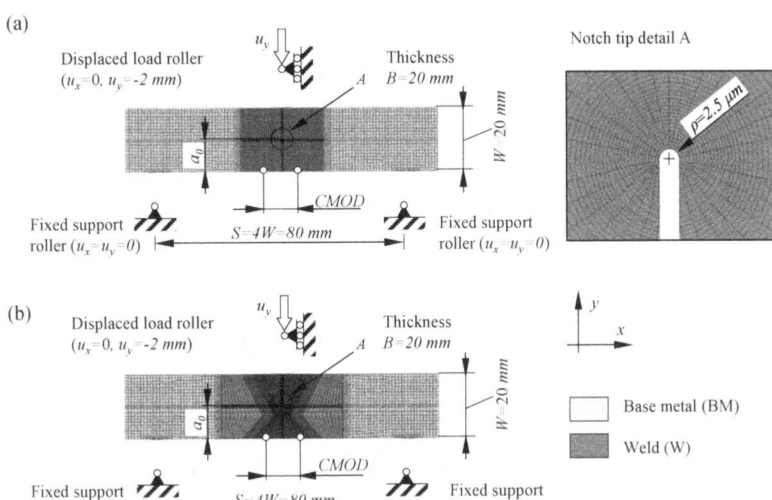

Figure 7. Examples of finite element models of SE(B) specimens: (**a**) containing only base metal and (**b**) containing weld.

Table 5. The analysis matrix with FEM series distinctive features.

Material	K_{Jlc} [MPa·m$^{1/2}$]	r_p [mm]
Base material (S690 QL)	69	1.6×10^{-3}
OM weld material (Mn4Ni2CrMo)	65	0.9×10^{-3}
UM weld material (G4Si1)	61	2.1×10^{-3}

The finite element mesh consisted of CPE8R eight node general purpose plane strain elements with reduced integration. In total, the mesh of the analyzed SE(B) models consisted of 10,580 to 10,810 finite elements, depending on crack length a_0/W and weld joint configuration. Symmetry conditions were not implemented so that the FEM models allow the replication of asymmetrical weld positions in future work.

Support and load rollers have been modelled as analytical rigid wire parts in the 2D plane strain finite element model. Boundary conditions have been prescribed in reference points at the center of each roller. Parametric studies included three different setups of support and load rollers, which are shown in Figure 8a–c. The first setup, which replicated the standard setup according to ASTM E1820, included support and load rollers with diameters of $d_L = 10$ mm and $d_S = 8$ mm, respectively. Support rollers are free to move in the horizontal direction and are fixed in the vertical direction. The second setup served as a control to investigate the influence of load roller diameter on computed *J*-integral values. It included the same support rollers as the previous setup but the load roller diameter increased to $d_L = 25$ mm. The third setup replicated the actual setup of support and load rollers used in fracture testing of SE(B) specimens. All rollers had diameter $d_L = d_S = 25$ mm and support rollers were fixed in vertical and horizontal directions. In all finite element models, the load was introduced in displacement control with a prescribed displacement of magnitude 2 mm to the load roller. Rotations of rollers were fixed in all three setups. Contacts have been established between rigid rollers and the deformable SE(B) specimen with a prescribed coefficient of friction $\mu = 0.1$.

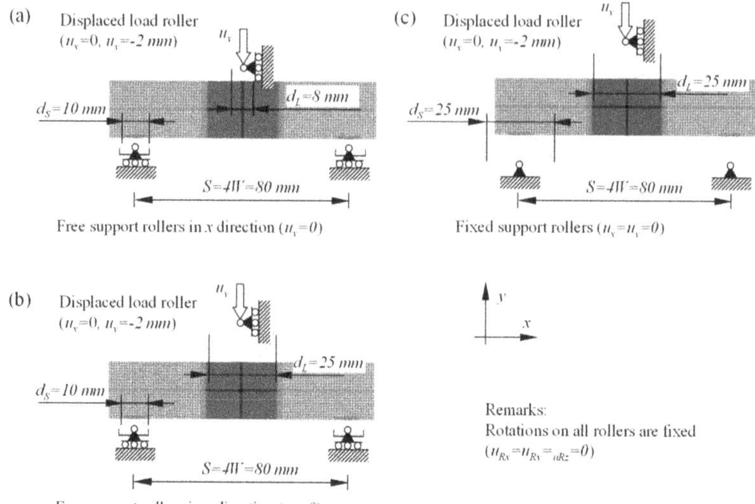

Figure 8. Overview of the investigated load and support rollers setups; (**a**) replica of standardized rollers setup according to ASTM E1820 standard, (**b**) control rollers setup and (**c**) replica of actual rollers setup used in fracture testing.

An elastic-plastic material model, which adopts J_2 flow theory with conventional von Mises plasticity, has been used to describe the material behavior under imposed loads. Material models of base, OM weld and UM weld metals have been determined on the basis of the experimental and offset stress–strain curves presented in Section 2. Plastic properties of listed materials have been implemented in FEM in form of true stress-true plastic strain curves which consisted of up to 21 data points. Finally, small strain assumptions have been implemented in order to enhance the J-integral convergence.

4. Evaluation of η_{pl} and γ_{pl} Factors

Calibration of the η_{pl} factor is based on a parametric plane strain elastic-plastic finite element analysis of SE(B) specimens with a stationary crack of various lengths. A method established by Donato et al. [15] is implemented in this work and it consists of the following steps:

1. From the results of FEA, extract applied load $P_{(i)}$, J-integral $J_{(i)}$, CMOD$_{(i)}$ and LLD$_{(i)}$ for each simulation increment. Here, J-integral is determined by a contour integral method.
2. Compute the area under the $P_{(i)}$-$V_{pl(i)}$ curve using Equation (7), where $V_{pl(i)}$ denotes plastic component of CMOD$_{(i)}$.
3. Compute the elastic component of the J-integral $J_{el(i)}$ using Equations (3)–(5).
4. Compute the plastic component of the J-integral $J_{pl(i)}$ using Equation (2). Here, $J_{(i)}$ is the J-integral extracted from the FEA results.
5. Compute the normalized area $A'_{pl(i)}$ under the $P_{(i)}$-$V_{pl(i)}$ curve by the following equation:

$$A'_{pl(i)} = \frac{A_{pl(i)}}{b^2 B \sigma_{YS}}.$$ (11)

6. Compute the normalized plastic component J'_{pl} of the J-integral by the following equation:

$$J'_{pl(i)} = \frac{J_{pl(i)}}{b \sigma_{YS}}$$ (12)

7. Create a plot of $J'_{pl(i)}$ as a function of $A'_{pl(i)}$. Compute the η_{pl} factor for the analysed SE(B) specimen as the slope of the $J'_{pl(i)}(A'_{pl(i)})$ plot by linear regression.

Once η_{pl} values are computed for SE(B) specimens with distinct normalized crack lengths a/W (in range $0.1 \leq a/W \leq 0.7$ in this work), the function $\eta_{pl}(a/W)$ can be determined by polynomial curve fitting of the results of η_{pl} as a function of a/W.

In the next step, γ_{pl} factor is evaluated, based on the framework established in the study of Zhu et al. [8], where functions η_{pl} and γ_{pl} of a/W are related as follows:

$$\gamma_{pl}(a/W) = \lambda\eta_{pl} - 1 - \frac{b}{W}\left(\frac{\lambda'}{\lambda} + \frac{\eta'_{pl}}{\eta_{pl}}\right). \tag{13}$$

Here, λ denotes a geometry function $\lambda(a/W)$, while λ' denotes the derivative $d\lambda(a/W)/d(a/W)$. Similarly, components η_{pl} and η'_{pl} denote the function $\eta_{pl}(a/W)$ and its derivative defined as $d\eta_{pl}(a/W)/d(a/W)$, respectively. The function $\lambda(a/W)$ is defined as the following ratio:

$$\lambda(a/W) = \frac{V_{pl}}{\Delta_{pl}}, \tag{14}$$

where V_{pl} and Δ_{pl} denote the plastic components of the CMOD and the LLD, respectively. The procedure for the evaluation of γ_{pl} has been established on the basis of Equations (15) and (16) and contains the following steps:

1. For each analyzed SE(B) specimen with distinct normalized crack length a/W, create a linear plot in which the plastic component of the LLD $\Delta_{pl(i)}$ is plotted as function of plastic component of the CMOD $V_{pl(i)}$. Compute the geometrical factor $\lambda_{(i)}$ as the slope of the $\Delta_{pl(i)}(V_{pl(i)})$ plot by linear regression method.
2. Determine the function $\lambda(a/W)$ by curve fitting of the FEA results of λ as a function of a/W in polynomial form.
3. Compute derivatives of $\lambda(a/W)$ and $\eta_{pl}(a/W)$ functions by differentiating them by the normalized crack length a/W:

$$\lambda'(a/W) = \frac{d\lambda(a/W)}{d(a/W)} \tag{15}$$

$$\eta'_{pl}\left(\frac{a}{W}\right) = \frac{d\eta_{pl}\left(\frac{a}{W}\right)}{d\left(\frac{a}{W}\right)}. \tag{16}$$

4. Compute values of η_{pl}, η'_{pl}, λ and λ' functions for each analyzed SE(B) specimen and insert them in Equation (12) in order to compute γ_{pl}.
5. Determine $\gamma_{pl}(a/W)$ by polynomial curve fitting of the computed results of γ_{pl} as a function of a/W.

5. Numerical Results of Plastic η_{pl} and γ_{pl} Factors

5.1. Verification of FEM and η_{pl} and γ_{pl} Factors for Base Material

Values of η_{pl} were obtained from the results of numerical simulations using the procedure described in Section 4. Values of J-integral for contours at 0.5 mm and 2.0 mm ahead of the crack tip (further referred as 0.5 mm contour and 2.0 mm contour respectively) were selected for the computation of the η_{pl} factors. The size of the 0.5 mm contour is relatively small and is suitable for the computation of J-integral in narrow double V welds, as the contour must be located in homogeneous material at the vicinity of the crack tip. The 2.0 mm contour has been used as the reference contour, since it provides converged values of the J-integral and does not interact with concentrated deformation due to load roller contact at a high load level in case of deep cracks with $a/W > 0.6$. Convergence analysis of the J contour integral was performed for the plane strain model of the base material SE(B) specimen with $a/W = 0.5$ and standard configuration of rollers at different load levels. The load levels are (normalized by the limit load F_y): $F/F_y = 0.53$, $F/F_y = 1$ and $F/F_y = 1.21$; the results are shown in Figure 9a–c respectively. The analyses showed that contours close to the crack tip ($y < 0.1$ mm) present inconsistent values of J-integral due to severe crack tip blunting and thus deformation of finite elements closest to the crack tip. Values of J integral fully converge at a distance 2.0 mm ahead of the crack tip with exception of high load levels exceeding $F/F_y = 1.21$. However, the relative deviation of J values for distant contours with respect to the 2.0 mm contour is less than 2%. Further comparison of J values for 0.5 mm and 2.0 mm contours showed the deviation between both to be less than 3.2% for all load levels, as presented in Figure 9d. Here, J values for the 2.0 mm contour represented reference values for computation of the relative deviation.

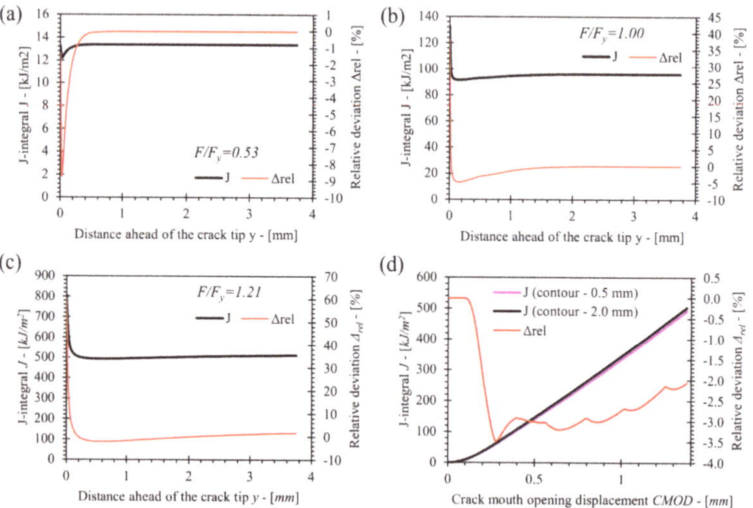

Figure 9. Results of J-integral convergence analyses at load levels (**a**) $F/F_y = 0.53$, (**b**) $F/F_y = 1$, (**c**) $F/F_y = 1.21$ and (**d**) comparison of obtained J-integral values for 0.5mm and 2.0 mm contours for monotonically loaded FEM of base material SE(B) with $a/W = 0.5$.

Verification of the developed finite element model has been conducted for SE(B) geometries consisting of base material only and implementing the standard configuration of load and support rollers (FEM series 1a). Obtained values of the η_{pl} factors for various crack lengths in the range $0.1 \leq a/W \leq 0.7$ were compared to values were published by Wu et al. [28], Kirk and Dodds [10], Nevalainen and Dodds [29], Kim and Schwalbe [11], Kim et al. [30], Donato and Ruggieri [12] and Zhu et al. [8]. Zhu et al. compared solutions proposed by the other listed references and provided solutions for η_{pl}, λ and γ_{pl}, that were eventually included in ASTM E1820, by curve fitting of the compared results. In this study, two sets of η_{pl} values, computed from the J values obtained from FEM along 0.5 mm and 2.0 mm contours, were included in the comparison. As demonstrated in Figure 9, both sets of η_{pl} values closely match the existing solutions. The former showed slightly increased deviations for cracks with normalized length $a/W < 0.3$ and $a/W > 0.5$, while the latter showed excellent agreement along the entire normalized crack length range $0.1 < a/W < 0.7$. This is due to the fact that convergence of J values improves with increasingly distant contours from the crack tip. The relative deviation from η_{pl} values included in standard ASTM E1820 is less than 5% in both cases. Additionally, it is important to emphasize that η_{pl} values produced in scope of this research do not exhibit increased variation for cracks with normalized length $a/W < 0.2$ as is the case for the solution obtained by Donato and Ruggieri [12], shown in Figure 10. The reason is that η_{pl} values in this paper were computed as slope of function $J'_{pl(i)}$ $(A'_{pl(i)})$ given by Equations (11) and (12) (presented in Section 4). In contrary, Donato and Ruggieri [12] computed η_{pl} values as an average of function $\eta_{pl}(CMOD)$ values that exceeded rigid exclusion condition $A_{pl(i)} \geq A_{(i)}$, where $A_{(i)}$ is area under P-CMOD curve. As a result, η_{pl} values that did not fully converge can be included in the computation of the average η_{pl} value for the given crack length, as shown in Figure 11b,c. Support of this discussion is shown in Figure 11a, where η_{pl} values computed from the created FEM by method of Donato and Ruggieri [12] follow the existing solution provided by Donato and Ruggieri. Based on the above stated arguments, it is assumed that the created FEM has passed the verification process and produces results that are in line with solutions from published researches and the standard ASTM E1820.

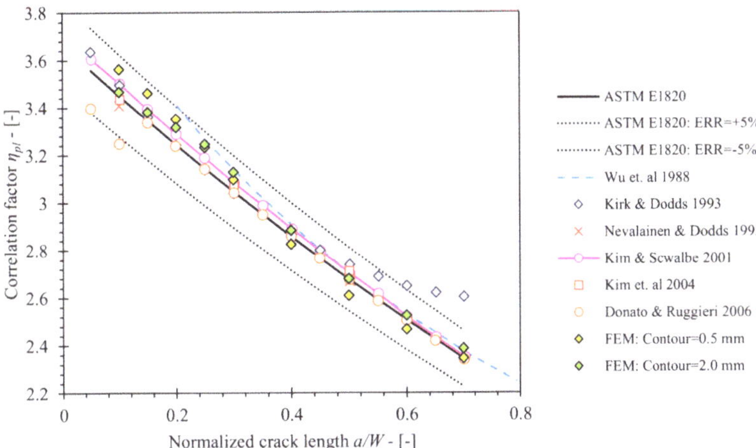

Figure 10. Comparison of η_{pl} values obtained from the developed FEM framework with the values published in the literature and standard ASTM E1820.

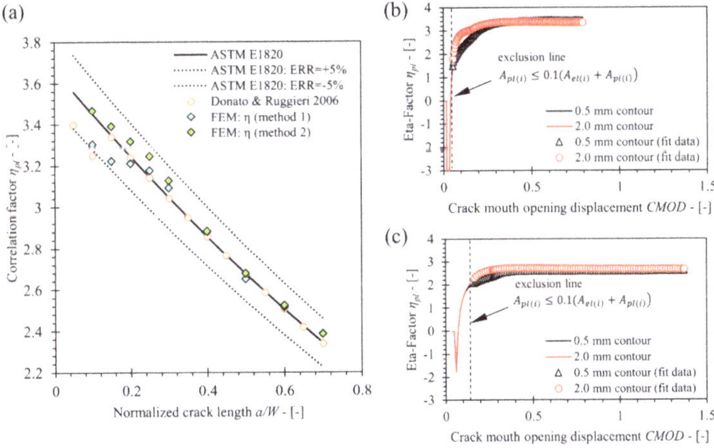

Figure 11. Comparison of (**a**) the current η_{pl} values computed by method described in Section 4 (denoted as method 1) and method developed by Donato and Ruggieri [12] (denoted as method 2) with the existing solutions. Influence of exclusion criterion on computation of η_{pl} according to the reference method is demonstrated for SE(B) specimens with crack length (**b**) $a/W = 0.15$ and (**c**) $a/W = 0.5$.

Following the FEM verification, remaining configurations of rollers 1b (standard setup, including oversized rollers with diameter $d_S = d_L = 25$ mm) and 1c (fixed oversized rollers with diameter $d_S = d_L = 25$ mm) were included in the investigation. Values of η_{pl}, computed from J-integral that was extracted from 0.5 mm and 2.0 mm contours, are presented in Figure 12. Here, the influence of the boundary conditions, that is, roller setup, on η_{pl} can be recognized.

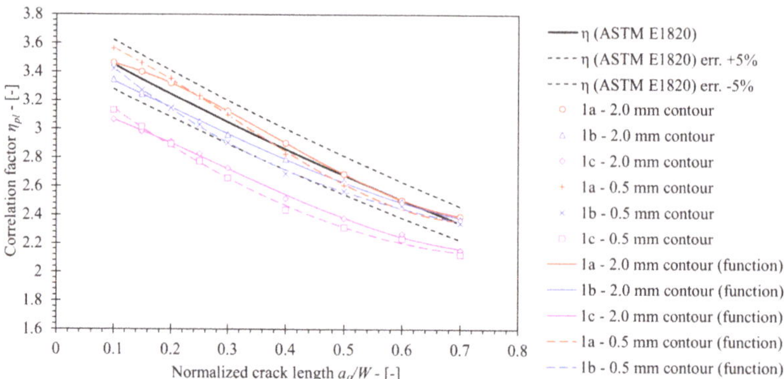

Figure 12. Comparison of η_{pl} values computed for base material using FEM with standard solution according to ASTM E1820. The former has been computed for numerical models of SE(B) specimens with various fixture and contour configurations.

First, if the standard diameter of the load roller $d_L = 10$ mm, implemented in FEM series 1a, is increased to $d_L = 25$ mm, control FEM series 1b, then the η_{pl} decreases at maximum 3.1% with respect to the standard solution (ASTM E1820) and 5.4% with respect to the baseline solution 1a, when J from the 2.0 mm contour is considered. However, both stated solutions seem to be in close agreement when the normalized crack length is $a/W \geq 0.6$.

Second, η_{pl} further decreases at maximum 12.0% with respect to the standard solution (ASTM E1820) and 13.3% with respect to the reference solution 1a in case of fixed load and support rollers with diameter $d_S = d_L = 25$ mm, implemented in FEM series 1c. Again, comparison of both solutions is based on the J-integral from the 2.0 mm contour.

Third, η_{pl} values computed on basis of J, extracted from the 0.5 mm contour, deviate from the values computed on basis of J that was obtained from 2.0 mm contour due to J not being fully converged. The former values deviate from the latter by 2.9%, 3.6% and 2.6 % at most for FEM series 1a, 1b and 1c, respectively.

Eventually, η_{pl} functions were developed by polynomial least squares curve fitting of the computational results. Proposed solutions are presented in Tables 6 and 7 for 0.5 mm and 2.0 mm contours, respectively.

Table 6. Proposed η_{pl} functions for SE(B) specimens of base material, valid for J, extracted from 0.5 mm contour.

Figure	η_{pl} Functions for 0.5 mm Contour in Range $0.1 \leq a/W \leq 0.7$	R^2 [-]
1a	$\eta_{pl} = 4.711\left(\frac{a}{W}\right)^3 - 4.339\left(\frac{a}{W}\right)^2 - 1.236\left(\frac{a}{W}\right) + 3.729$	0.999
1b	$\eta_{pl} = -1.121\left(\frac{a}{W}\right)^3 + 3.300\left(\frac{a}{W}\right)^2 - 3.789\left(\frac{a}{W}\right) + 3.774$	0.999
1c	$\eta_{pl} = 0.296\left(\frac{a}{W}\right)^3 + 1.556\left(\frac{a}{W}\right)^2 - 3.094\left(\frac{a}{W}\right) + 3.437$	0.998

Table 7. Proposed η_{pl} functions for SE(B) specimens of base material, valid for *J*, extracted from 2.0 mm contour.

FEM Series	η_{pl} Functions for 2.0 mm Contour in Range $0.1 \leq a/W \leq 0.7$	R^2 [-]
1a	$\eta_{pl} = 5.025\left(\frac{a}{W}\right)^3 - 5.738\left(\frac{a}{W}\right)^2 - 0.061\left(\frac{a}{W}\right) + 3.552$	0.999
1b	$\eta_{pl} = 1.226\left(\frac{a}{W}\right)^3 - 0.684\left(\frac{a}{W}\right)^2 - 1.751\left(\frac{a}{W}\right) + 3.519$	0.998
1c	$\eta_{pl} = 2.483\left(\frac{a}{W}\right)^3 - 2.181\left(\frac{a}{W}\right)^2 - 1.183\left(\frac{a}{W}\right) + 3.206$	0.998

The geometry factor λ has been determined on the basis of LLD and CMOD according to Equation (14). Figure 13 presents obtained values of the geometry factor λ as a function of a/W. Figure 13 demonstrates that λ values, computed for FEM series 1a (standard roller setup) are in close agreement with solution included in ASTM E1820 for $a/W \geq 0.25$. In case of shallower cracks with $a/W < 0.25$, computed results deviate from the standard solution. The reason is that Zhu et al. [8] produced the solution for λ by curve fitting of the existing results in the range $0.25 \leq a/W \leq 0.7$. Further inspection of computed results shows that values of λ increase by 8.1% at maximum with respect to the reference solution for FEM series 1a if the load roller diameter is increased from 8 mm to 25 mm (FEM series 1b). However, increasing the support rollers diameter from 10 mm to 25 mm and constraining their degrees of freedom has little effect on λ values in case of FEM series 1c as computed values deviate 10.9% at maximum with respect to the reference solution for FEM series 1a. Finally, λ functions were developed by polynomial curve fitting of the computed results, using the least squares method. Proposed solutions are presented in Table 8.

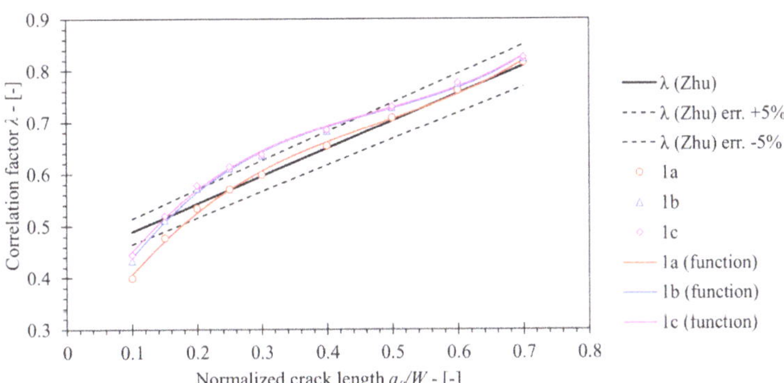

Figure 13. Comparison of λ values computed by FEM for base material with standard solution according to ASTM E1820 that was originally proposed by Zhu [8].

Table 8. Proposed λ functions for SE(B) specimens of base material.

FEM Series	λ Functions in Range $0.1 \leq a/W \leq 0.7$	R^2 [-]
1a	$\lambda = 2.278\left(\frac{a}{W}\right)^3 - 3.273\left(\frac{a}{W}\right)^2 + 2.008\left(\frac{a}{W}\right) + 0.236$	0.998
1b	$\lambda = 2.806\left(\frac{a}{W}\right)^3 - 4.001\left(\frac{a}{W}\right)^2 + 2.249\left(\frac{a}{W}\right) + 0.252$	0.998
1c	$\lambda = 2.603\left(\frac{a}{W}\right)^3 - 3.707\left(\frac{a}{W}\right)^2 + 2.113\left(\frac{a}{W}\right) + 0.274$	0.997

The crack growth correction factor γ has been determined from computed solutions for η_{pl} and λ functions and their derivatives according to Equation (14). A significant deviation of the computed γ value functions from the standard solution is observed (Figure 14). Solutions obtained from FEM series 1c and 1a deviate from the standard solution by a factor of 6.1 to 7.7 at most respectively, where *J* was evaluated from the 2.0 mm contour.

Similarly, solutions based on *J*, evaluated from the 0.5 mm contour, deviated from the standard solution by a factor of 7.2 to 8.4 at most. In both cases, solutions obtained from FEM series 1b deviated from the standard solution within the specified ranges. Revision of post-processing procedures revealed that such deviations are due to the combination of η_{pl}, η_{pl}', λ and λ' values in Equation (14). The obtained solutions are presented in Tables 9 and 10 for *J* evaluated from 0.5 mm and 2.0 mm contours.

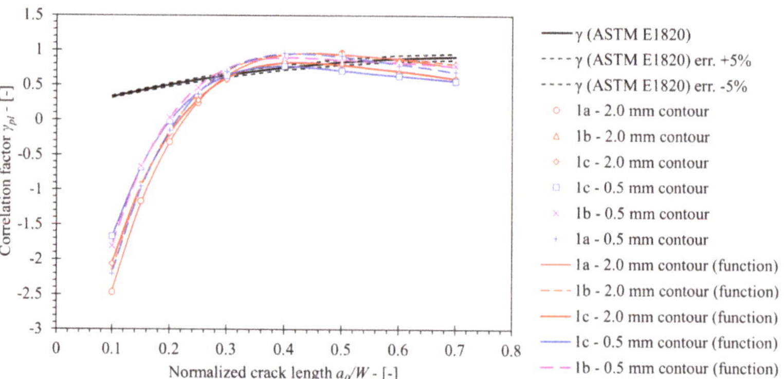

Figure 14. Comparison of the crack growth correction factor γ computed for base material using FEM with standard solution according to ASTM E1820. The former has been computed for numerical models of SE(B) specimens with various fixture and contour configurations.

Table 9. Proposed γ functions for SE(B) specimens of base material, valid for *J* extracted from 0.5 mm contour.

FEM Series	γ Functions for 0.5 mm Contour in Range $0.1 \leq a/W \leq 0.7$	R^2 [-]
1a	$\gamma_{pl} = -65.073\left(\frac{a}{W}\right)^4 + 152.019\left(\frac{a}{W}\right)^3 - 132.906\left(\frac{a}{W}\right)^2 + 50.489\left(\frac{a}{W}\right) - 6.048$	1.000
1b	$\gamma_{pl} = -77.924\left(\frac{a}{W}\right)^4 + 170.124\left(\frac{a}{W}\right)^3 - 137.863\left(\frac{a}{W}\right)^2 + 48.739\left(\frac{a}{W}\right) - 5.440$	0.999
1c	$\gamma_{pl} = -63.999\left(\frac{a}{W}\right)^4 + 144.314\left(\frac{a}{W}\right)^3 - 120.653\left(\frac{a}{W}\right)^2 + 43.576\left(\frac{a}{W}\right) - 4.955$	0.999

Table 10. Proposed γ functions for SE(B) specimens of base material, valid for *J* extracted from 2.0 mm contour.

FEM Series	γ Functions for 2.0 mm Contour in Range $0.1 \leq a/W \leq 0.7$	R^2 [-]
1a	$\gamma_{pl} = -65.574\left(\frac{a}{W}\right)^4 + 152.247\left(\frac{a}{W}\right)^3 - 133.564\left(\frac{a}{W}\right)^2 + 51.641\left(\frac{a}{W}\right) - 6.420$	1.000
1b	$\gamma_{pl} = -73.362\left(\frac{a}{W}\right)^4 + 164.154\left(\frac{a}{W}\right)^3 - 137.889\left(\frac{a}{W}\right)^2 + 51.007\left(\frac{a}{W}\right) - 6.032$	1.000
1c	$\gamma_{pl} = -60.373\left(\frac{a}{W}\right)^4 + 139.811\left(\frac{a}{W}\right)^3 - 121.465\left(\frac{a}{W}\right)^2 + 46.027\left(\frac{a}{W}\right) - 5.563$	1.000

5.2. η_{pl} and γ_{pl} Factors for OM and UM Welds

A postprocessing procedure similar to the one described in the previous paragraph has been applied to results of finite element analyses of welded SE(B) specimens. The FEM model of the welded SE(B) specimens is described in detail in Section 3.2. It is important to emphasize that in this case values of *J* integral were obtained from the 0.5 mm contour only. This compact contour can be entirely located in weld material when the crack tip is located in a narrow weld root, thus meeting the requirements of material homogeneity for computation of the *J* contour integral. One disadvantage is that values of *J*-integral are not fully converged at a distance 0.5 mm ahead of the crack tip. However, they deviate less than 3.2% in comparison with J values obtained from the 2.0 mm contour, as reported in previous paragraph, which is considered acceptable for the following.

Additionally, the effect of weld yield strength mismatch variation on fracture behavior of the analyzed SE(B) specimens has been investigated. The actual produced OM and UM weld materials have mismatch ratio $M = 1.31$ and $M = 0.78$ with respect to the base material S690 QL, respectively. Additional mismatch ratios $M = 1.5$ and $M = 0.5$ were investigated by implementing material models of the OM and UM weld materials that were obtained by offsetting true tress-strain curves, as described in Section 2.2.

Again, values of η_{pl} were computed by the Eta-method described in Section 2.2 and are represented in Figure 15. All simulations implemented a fixed roller setup, which replicated the boundary conditions of the actual performed fracture toughness tests. The computed solution for the base material is plotted as the reference; several observations can be made on basis of the plotted results.

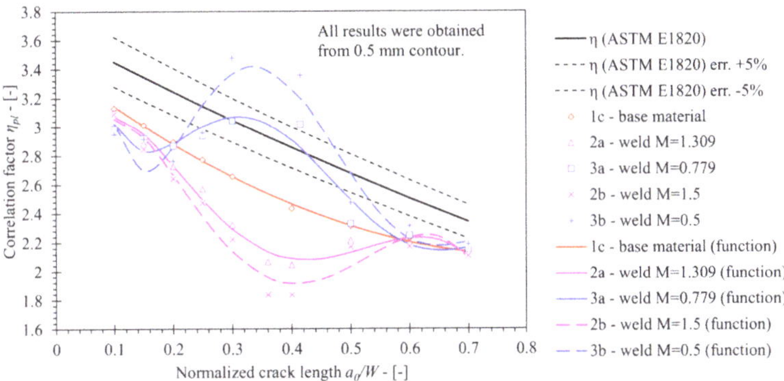

Figure 15. Comparison of η_{pl} values computed for weld material using FEM with standard solution according to ASTM E1820. The former has been computed for weld materials with various mismatch factors M. Solution, obtained for base material, has been plotted as a reference.

First, the influence of the weld geometry and its mechanical properties on η_{pl} is clearly demonstrated. Values of η_{pl} decrease significantly when the crack length a/W is similar to the distance to the weld root $L_W/W = 0.36$ in case of the OM weld material. The opposite can be observed for UM weld material, where $L_W/W = 0.42$.

Second, altered material models produce η_{pl} values with pronounced minimum and maximum values when a/W is similar to L_W/W in case of yield strength mismatch $M = 0.5$ and $M = 1.5$ respectively. Computed η_{pl} values are 12.2% higher at $a/W = 0.36$ if an altered material model with $M = 0.5$ is implemented in FEM instead of the actual UM material. Moreover, computed η_{pl} values are 6.9% lower at $a/W = 0.415$ if an altered material model with $M = 1.5$ is implemented in FEM instead of the actual OM material.

Functions of η_{pl} were obtained by polynomial least squares curve fitting of the computed results. The degree of fitted polynomial functions has been carefully selected in order to improve the coefficient of regression R^2 while avoiding excessive variations that are characteristic for higher degree polynomials. The proposed solutions are presented in Table 11.

Table 11. Proposed η_{pl} functions for investigated actual and altered weld material SE(B) specimens.

FEM Series	η_{pl} Functions for 0.5 mm Contour in Range $0.1 \leq a/W \leq 0.7$	R^2 [-]
2a	$\eta_{pl} = 92.933\left(\frac{a}{W}\right)^5 - 269.918\left(\frac{a}{W}\right)^4 + 264.133\left(\frac{a}{W}\right)^3 - 104.146\left(\frac{a}{W}\right)^2 + 13.094\left(\frac{a}{W}\right) + 2.569$	0.979
2b	$\eta_{pl} = 97.960\left(\frac{a}{W}\right)^5 - 316.782\left(\frac{a}{W}\right)^4 + 326.940\left(\frac{a}{W}\right)^3 - 132.696\left(\frac{a}{W}\right)^2 + 17.477\left(\frac{a}{W}\right) + 2.339$	0.957
3a	$\eta_{pl} = -380.531\left(\frac{a}{W}\right)^5 + 877.582\left(\frac{a}{W}\right)^4 - 739.216\left(\frac{a}{W}\right)^3 + 274.115\left(\frac{a}{W}\right)^2 - 43.820\left(\frac{a}{W}\right) + 5.321$	0.934
3b	$\eta_{pl} = -715.044\left(\frac{a}{W}\right)^5 + 1647.962\left(\frac{a}{W}\right)^4 - 1395.160\left(\frac{a}{W}\right)^3 + 522.880\left(\frac{a}{W}\right)^2 - 83.327\left(\frac{a}{W}\right) + 7.365$	0.880

Following the analysis of η_{pl}, the geometry factor λ has been computed by inserting CMOD and LLD from FEM results into Equation (14). Figure 16 presents computed solutions, which are in close agreement for deep cracked SE(B) specimens with $a/W \geq 0.5$. Furthermore, the influence of material properties is demonstrated for shallower cracks. Here, values of λ are lower for welded joints in comparison with all-base metal specimens, while OM weld configurations produce higher values of λ in comparison with UM weld configurations. Computed values of λ for OM and UM weld configurations deviate from the one of the base material by 13.5% and 19.1% at most, respectively. However, welds with altered yield strength mismatch $M = 1.5$ and $M = 0.5$ exhibit lower values of λ in comparison to the OM weld with $M = 1.31$ by 11.5% and the UM weld with $M = 0.779$ by 10.9% at most, respectively. Finally, λ functions that are presented in Table 12, were developed by the least squares method on the basis of the aforementioned results.

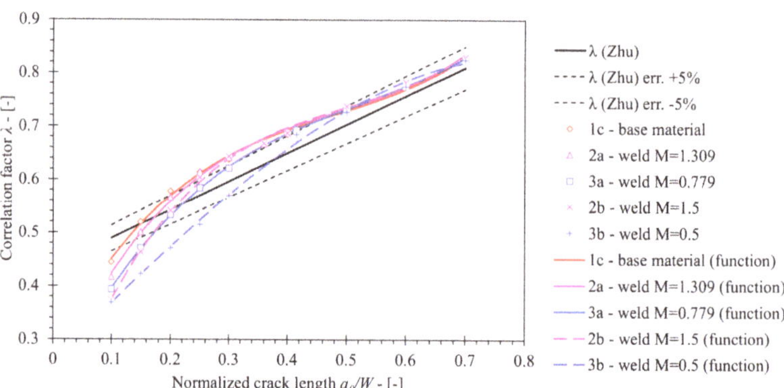

Figure 16. Comparison of λ values computed by FEM for weld material with standard solution according to ASTM E1820 that was originally proposed by Zhu [8]. The former has been computed for welds with various mismatch factors M. Solution obtained for base material has been plotted as a reference.

Table 12. Proposed λ_l functions for investigated actual and altered weld material SE(B) specimens.

FEM Series	λ Functions for 0.5 mm Contour J-Integral in Range $0.1 \leq a/W \leq 0.7$	R^2 [-]
2a	$\lambda = 3.000\left(\frac{a}{W}\right)^3 - 4.333\left(\frac{a}{W}\right)^2 + 2.442\left(\frac{a}{W}\right) + 0.220$	0.997
2b	$\lambda = 3.547\left(\frac{a}{W}\right)^3 - 5.285\left(\frac{a}{W}\right)^2 + 2.972\left(\frac{a}{W}\right) + 0.127$	0.998
3a	$\lambda = 2.229\left(\frac{a}{W}\right)^3 - 3.505\left(\frac{a}{W}\right)^2 + 2.254\left(\frac{a}{W}\right) + 0.203$	1.000
3b	$\lambda = -0.469\left(\frac{a}{W}\right)^3 - 0.141\left(\frac{a}{W}\right)^2 + 1.135\left(\frac{a}{W}\right) + 0.256$	0.998

Finally, solutions for the crack growth correction factor γ have been obtained by Equation (13) and are presented in Figure 17, where the solution for the base material is plotted as a reference. Here, the γ factors for OM and UM weld configurations vary significantly throughout the range $0.1 \leq a/W \leq 0.7$. The analysis of η_{pl} and λ factors and their derivatives reveal that, in the case of welds, the shape of the η_{pl} function has a significant impact on the variation of the γ function. The computed γ functions for OM and UM weld configurations deviate from the base material solution by a factor of 10 and 5.6 at most, respectively. Furthermore, the γ factor for welds with altered yield strength mismatch $M = 1.5$ and $M = 0.5$ deviate in comparison with the OM weld configuration with $M = 1.31$ by a factor of 9.0 and the UM weld configuration with $M = 0.779$ by a factor of 17.5 at most, respectively. Finally, γ_{pl} functions that are presented in Table 13 were developed by the least squares method on the basis of the aforementioned results.

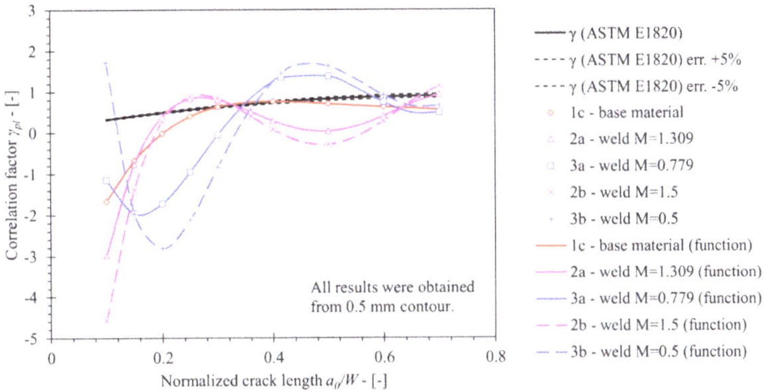

Figure 17. Comparison of the crack growth correction factor γ computed for weld material using FEM with standard solution according to ASTM E1820. The former has been computed for welds with various mismatch factors M. Solution obtained for the base material has been plotted as a reference.

Table 13. Proposed γ_{pl} functions for investigated actual and altered weld material SE(B) specimens.

FEM Series	γ_{pl} Functions for 0.5 mm Contour in Range $0.1 \leq a/W \leq 0.7$	R^2 [-]
2a	$\gamma_{pl} = -350.776\left(\frac{a}{W}\right)^4 + 691.443\left(\frac{a}{W}\right)^3 - 476.044\left(\frac{a}{W}\right)^2 + 133.556\left(\frac{a}{W}\right) - 12.237$	1.000
2b	$\gamma_{pl} = -501.441\left(\frac{a}{W}\right)^4 + 992.155\left(\frac{a}{W}\right)^3 - 683.944\left(\frac{a}{W}\right)^2 + 191.723\left(\frac{a}{W}\right) - 17.844$	1.000
3a	$\gamma_{pl} = -451.632\left(\frac{a}{W}\right)^5 + 1422.106\left(\frac{a}{W}\right)^4 - 1562.720\left(\frac{a}{W}\right)^3 + 734.779\left(\frac{a}{W}\right)^2 - 136.680\left(\frac{a}{W}\right) + 6.601$	1.000
3b	$\gamma_{pl} = -1503.250\left(\frac{a}{W}\right)^5 + 4096.377\left(\frac{a}{W}\right)^4 - 4175.130\left(\frac{a}{W}\right)^3 + 1940.616\left(\frac{a}{W}\right)^2 - 391.926\left(\frac{a}{W}\right) + 25.285$	0.999

6. Application to Fracture Toughness Testing

Fracture toughness tests using the single specimen method according to standard ASTM E1820 were performed as described in Section 2.3. J-integral resistance (i.e., J-R) curves for base material, OM weld and UM weld have been computed using the equations for factors η_{pl}, and γ_{pl}, numerically evaluated in the scope of this research (Section 5). For the purpose of comparison, equations for factors η_{pl}, and γ_{pl}, included in standard ASTM E1820 were used to produce reference J-R solutions for the base material and both weld configurations. Computed J-R curves and relative errors with respect to standard J-R solutions are shown in Figures 18–21. Values of the J-integral at crack growth onset J_{Ic} were determined at the intersection of the J-R curves with the 0.2 mm blunting line and are presented in Tables 14–16.

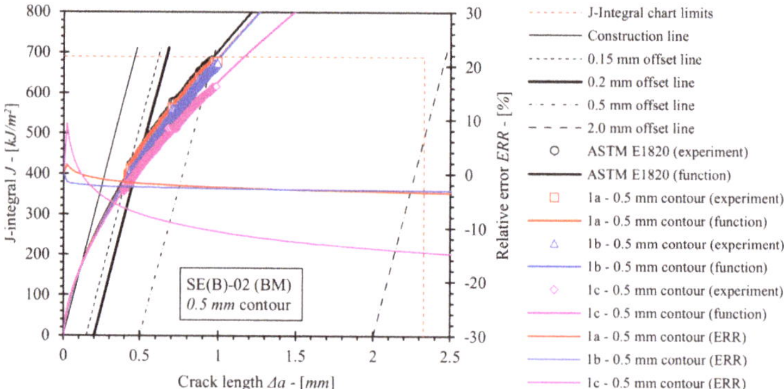

Figure 18. Evaluated *J-R* curves for base material, where η_{pl} and γ_{pl} functions, evaluated on the basis of numerical results, obtained from the 0.5 mm contour, were implemented. *J-R* curves, computed by ASTM E1820 are plotted as a reference.

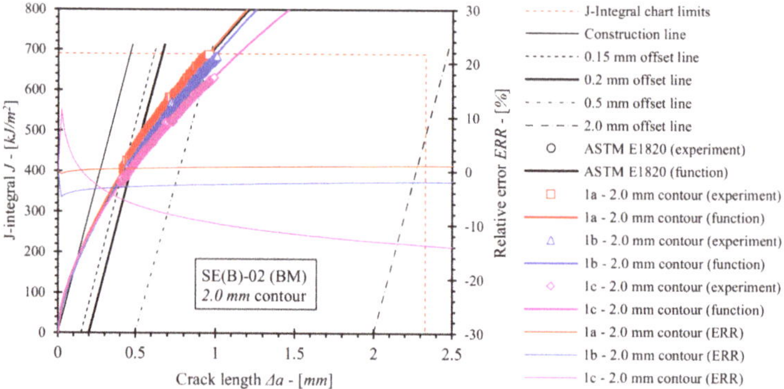

Figure 19. Evaluated *J-R* curves for base material, where η_{pl} and γ_{pl} functions, evaluated on the basis of numerical results, obtained from the 2.0 mm contour, were implemented. *J-R* curves, computed by ASTM E1820 are plotted as a reference.

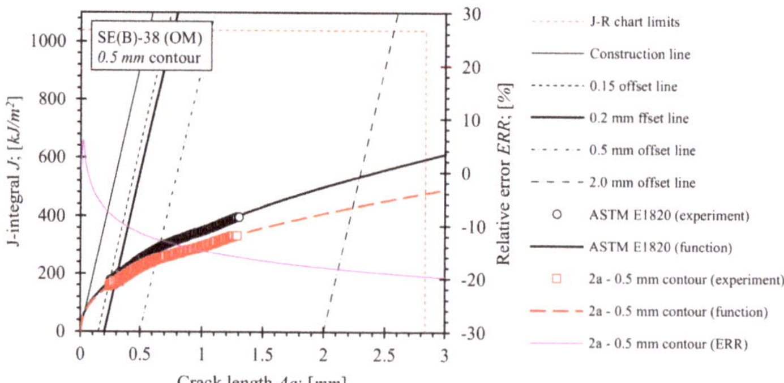

Figure 20. Evaluated *J-R* curves for the OM weld. η_{pl} and γ_{pl} functions, evaluated on the basis of numerical results, obtained from the 0.5 mm contour, were implemented. *J-R* curves that were computed by ASTM E1820 are plotted as a reference.

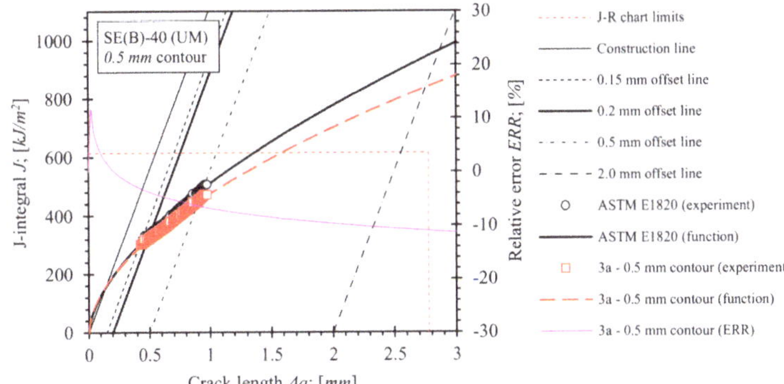

Figure 21. Evaluated *J*-*R* curves for the UM weld. η_{pl} and γ_{pl} functions, evaluated on basis of numerical results, obtained from the 0.5 mm contour, were implemented. *J*-*R* curves that were computed by ASTM E1820 are plotted as a reference.

Table 14. Computed results of fracture toughness testing of the base material.

Basic Specimen Data	Computation Case, Contour	J_{Ic} [kJ/m²]	K_{JIc} [MPa·m$^{1/2}$]
	ASTM E1820	448	313
Designation: SE(B)-02	1a, 0.5 mm	437	309
Material: base material (S690 QL)	1a, 2.0 mm	456	315
a_0 = 10.645 mm a_0/W = 0.533	1b, 0.5 mm	421	303
a_f = 11.785 mm Δa = 1.140 mm	1b, 2.0 mm	430	306
	1c, 0.5 mm	400	295
	1c, 2.0 mm	408	298

Table 15. Computed results of fracture toughness testing of the overmatched weld material.

Basic Specimen Data	Computation Case, Contour	J_{Ic} [kJ/m²]	K_{JIc} [MPa·m$^{1/2}$]
Designation: SE(B)-38	ASTM E1820	193	224
Material: OM weld material (Mn4Ni2CrMo)			
a_0 = 8.579 mm a_0/W = 0.431	2a, 0.5 mm	173	212
a_f = 9.888 mm Δa = 1.309 mm			

Table 16. Computed results of fracture toughness testing of the undermatched weld material.

Basic Specimen Data	Computation Case, Contour	J_{Ic} [kJ/m²]	K_{JIc} [MPa·m$^{1/2}$]
Designation: SE(B)-40	ASTM E1820	336	294
Material: UM weld material (G4Si1)			
a_0 = 9.015 mm a_0/W = 0.448	1a, 0.5 mm	318	285
a_f = 9.995 mm Δa = 0.980 mm			

It should be noted that *J*-*R* curves for the base material (Figures 18 and 19), computed with η_{pl} and γ_{pl} that were calibrated on basis of FEM series 1a results, are in agreement with standard *J*-*R* solutions. Here, deviations are less than 4% for the 0.5 mm contour (Figure 18) and less than 2% for the 2.0 mm contour (Figure 19). Similarly, values of J_{Ic} deviate up to 2.5% and 1.8% with respect to standard solutions for *J*-integral computed by 0.5 mm and

2.0 mm contours, respectively. This indicates that the FEM framework has been properly constructed. Moreover, various boundary conditions (i.e., load and support rollers setup) have an effect on *J-R* solutions. If the diameter of the load roller is increased from 8 mm to 25 mm as in computation case 1b, then the following observations can be made. Values of the *J* integral throughout the total crack extension Δa are comparable to solution 1a if η_{pl} and γ_{pl}, valid for the 0.5 mm contour *J*-integral are implemented. In contrast, values of *J* decrease up to 4% with respect to standard *J-R* solution in case of η_{pl} and γ_{pl}, valid for the 2.0 mm contour. Finally, if support rollers are constrained and have diameter increased from 10 mm to 25 mm as in computation case 1c, values of *J*-integral throughout the total crack extension Δa decrease up to 15% and 14% with respect to the standard *J-R* solution if η_{pl} and γ_{pl}, valid for 0.5 mm and 2.0 mm contours are implemented respectively. Moreover, J_{Ic} is decreased by 10.7% and 8.9% with respect to the standard solution for the 0.5 mm and 2.0 mm contours respectively. All in all, *J* is overestimated if the standard computational procedure is utilized to postprocess results of a fracture toughness test where the modified roller setup has been implemented.

Similar observations can be made for fracture toughness test results of OM and UM welds, shown in Figures 20 and 21 respectively. Here, the presented *J-R* curves were computed for load and fully constrained support rollers with diameter of 25 mm, while η_{pl} and γ_{pl} functions, calibrated on the basis of the results of FEM series 2a and 3a for the 0.5 mm contour, have been implemented in the computation. Comparison of *J-R* curves based on numerical results with the standard *J-R* solutions reveals that *J* values throughout the total crack extension Δa are decreased up to 20% for the OM weld (Figure 20) and up to 10% for the UM weld (Figure 21). Moreover, the corresponding J_{Ic} values, listed in Tables 15 and 16, indicate a reduction by 10.4% and 5.4% for OM and UM welds, respectively. Again, *J* values are overestimated if the standard computational procedure is used for the postprocessing of results of the fracture toughness test with the utilized modified roller setup.

7. Conclusions

This work addresses the effect of weld strength mismatch and configuration of fixtures on *J* estimation formulas to determine fracture toughness from laboratory measurements using the load-CMOD data of SE(B) specimens. The investigation included fracture toughness testing of standard base material SE(B) specimens and surface cracked SE(B) specimens containing an overmatching or undermatching weld, where the notch and welded joint were aligned with the central plane. A new set of η_{pl}, λ and γ_{pl} factors for computing *J* from experimental results has been developed for a wide range of crack sizes ($0.1 \leq a/W \leq 0.7$), levels of weld yield strength mismatch and fixture configurations. Aforementioned factors were developed on the basis of computational results, provided by parametric 2D plane strain finite element analyses. The major conclusions of this study can be summarized as follows:

- Varying the configuration of fixtures has an effect on η_{pl}, λ and γ_{pl} factors as demonstrated by parametric plane strain analyses of the base material SE(B) specimen. Values of η_{pl} decrease by 6.4% if the load roller diameter is increased from $d_L = 8$ mm to $d_S = 25$ mm and by 14.2% if load and support rollers diameter is set to $d_L = d_S = 25$ mm, while the latter are fully constrained. Values of λ increase throughout the given range of crack lengths for both fixture modifications. Therefore, it is assumed that CMOD increases for the same LLD, while less work of the applied load manifests in the crack driving force if the fixture is modified.
- Weld geometry in conjunction with weld material properties has a direct effect on η_{pl} values. A distinctive decrease of η_{pl} values can be observed if the crack tip is near the weld root in case of an overmatching weld with $M = 1.31$, while the opposite can be observed in the case of an undermatching weld with $M = 0.78$. Modifying the mismatch to $M = 1.50$ (overmatching weld) and $M = 0.50$ (undermatching weld) further enhances the deviation of η_{pl} near the weld root. Moreover, weld material properties affect the λ factor, where all values are in close agreement when $a/W > 0.35$

with the exception of λ for modified undermatching weld material with $M = 0.50$. For $a/W < 0.35$ the λ factors for overmatching and undermatching (original and modified) weld material decrease in comparison with the λ for the base material. As a consequence of distinctive shapes of the η_{pl} and λ functions, the γ functions oscillate with respect to the standard solution.

- The computed resistances to stable tearing of the tested SE(B) specimens, expressed in terms of J-R curves, are highly impacted by the η_{pl}, λ and γ_{pl} factors. Values of J throughout the total crack extension Δa reduce with respect to the standard solution by 15% for the base material, by 11% for the undermatching weld and by 18% for the overmatching weld when the set of correction factors calibrated for the utilized fixture ($d_L = d_S = 25$ mm, constrained support rollers) is used. This indicates that fracture toughness can be overestimated if the standard η_{pl}, λ and γ_{pl} factors are applied to the postprocessing of results obtained by fracture toughness testing, where a modified fixture has been utilized.

Author Contributions: The research was conceptualized and supervised by N.G. and S.H. Experimental testing was performed by P.Š. and N.G. Investigative support and numerical analysis in frame of project were performed by P.Š. and S.N. Formal analysis and methodology were provided by W.D.W. All authors have read and agreed to the published version of the manuscript.

Funding: The authors would like to acknowledge ARRS (Slovenian Research Agency, research grant nr. N2-0030) and FWO Vlaanderen (Research Foundation—Flanders, research grant nr. G.0609.15N) for the support provided during this research.

Institutional Review Board Statement: Not applicable.

Informed Consent Statement: Not applicable.

Data Availability Statement: Not applicable.

Acknowledgments: The authors would like to acknowledge the Slovenian Research Agency (ARRS) for financial support of the Research Program P2-0137 "Numerical and experimental analysis of mechanical systems".

Conflicts of Interest: The authors declare no conflict of interest.

References

1. *BS 7910:2013*; Guide to Methods for Assessing the Acceptability of Flaws in Metallic Structures. BSI Standards Limited, Chiswick Tower: London, UK, 2015.
2. *Assessment of the Integrity of Structures Containing Defects*; R6, Revision 4; EDF Energy Nuclear Generation Ltd.: Barnwood, UK, 2001.
3. Koçak, M.; Webster, S.; Janosch, J.J.; Ainsworth, R.A.; Koers, R. *FITNET Fitness-for-Service (FFS)—Procedure*; GKSS Research Center: Geesthacht, Germany, 2008; Volume 1.
4. Koçak, M.; Hadley, I.; Szavai, S.; Tkach, Y.; Taylor, N. *FITNET Fitness-for-Service (FSS)—Annex*; GKSS Research Center: Geesthacht, Germany, 2008; Volume 2.
5. *ISO 15653:2010*; Metallic Materials—Method of Test for the Determination of Quasistatic Fracture Toughness of Welds. International Organization for Standardization: Geneva, Switzerland, 2010.
6. *BS 7448-2:1997*; Fracture Mechanics Toughness Test, Part-2. Method for Determination of KIc, Critical CTOD and Critical J Values of Welds in Metallic Materials. BSI Standards Limited, Chiswick Tower: London, UK, 2004.
7. *ASTM E1820-15*; Standard Test Method for Measurement of Fracture Toughness. ASTM International: West Conshohocken, PA, USA, 2016.
8. Zhu, X.K.; Leis, B.N.; Joyce, J.A. Experimental Estimation of J-R Curves from Load-CMOD Record for SE(B) Specimens. *J. ASTM Int.* **2008**, *5*, 1–15. [CrossRef]
9. Hellmann, D.; Rohwerder, G.; Schwalbe, K.H. Development of a Test Setup for Measuring the Deflection of Single-Edge Notched Bend (SENB) Specimens. *J. Test Eval.* **1984**, *12*, 42–44. [CrossRef]
10. Kirk, M.T.; Dodds, R.H. J and CTOD Estimation Equations for Shallow Cracks in Single Edge Notch Bend Specimens. *J. Test Eval.* **1993**, *21*, 228. [CrossRef]
11. Kim, Y.J.; Schwalbe, K.H. On Experimental J Estimation Equations Based on CMOD for SE(B) Specimens. *J. Test. Eval.* **2001**, *29*, 67–71. [CrossRef]

12. Donato, G.H.B.; Ruggieri, C. Estimation Procedures for *J* and CTOD Fracture Parameters Using Three-Point Bend Specimens. In Proceedings of the International Pipeline Conference, Calgary, AB, Canada, 25 September 2006; pp. 149–157.
13. Kim, Y.J.; Kim, J.S.; Schwalbe, K.H.; Kim, Y.J. Numerical investigation on *J*-integral testing of heterogeneous fracture toughness testing specimens: Part I—Weld metal cracks. *Fatigue Fract. Eng. Mater. Struct.* **2003**, *26*, 683–694. [CrossRef]
14. Eripret, C.; Hornet, P. Fracture toughness testing procedures for strength mis-matched structures. In *Mis-Matching of Interfaces and Welds*; Schwalbe, K.H., Koçak, M., Eds.; GKSS Research Center: Geestchacht, Germany, 1997.
15. Donato, G.H.B.; Magnabosco, R.; Ruggieri, C. Effects of weld strength mismatch on *J* and CTOD estimation procedure for SE(B) specimens. *Int. J. Fract.* **2009**, *159*, 1–20. [CrossRef]
16. Mathias, L.L.; Sarzosa, D.F.; Ruggieri, C. Effects of specimen geometry and loading mode on crack growth resistance curves of a high-strength pipeline girth weld. *Int. J. Press. Vessel. Pip.* **2013**, *111*, 106–119. [CrossRef]
17. *DIN EN ISO 15792-1:2012*; Schweißzusätze—Prüfverfahren—Teil 1: Prüfverfahren für Prüfstücke zur Entnahme von Schweißgutproben an Stahl, Nickel und Nickellegierungen. Deutsches Institut für Normung: Berlin, Germany, 2012.
18. *ASTM E8/E8M-13a*; Standard Test Methods for Tension Testing of Metallic Materials. ASTM International: West Conshohocken, PA, USA, 2013.
19. Schwalbe, K.H.; Heerens, J.; Zerbst, U.; Pisarski, H.; Koçak, M. *EFAM GTP 02—The GKSS Test Procedure for Determining the Fracture Behaviour of Materials*; GKSS Research Center: Geesthacht, Germany, 2002.
20. Landes, J.; Zhou, Z.; Lee, K.; Herrera, R. Normalization Method for Developing *J-R* Curves with the LMN Function. *J. Test. Eval.* **1991**, *19*, 305–311. [CrossRef]
21. Tang, J.; Liu, Z.; Shi, S.; Chen, X. Evaluation of fracture toughness in different regions of weld joints using unloading compliance and normalization method. *Eng. Fract. Mech.* **2018**, *195*, 1–12. [CrossRef]
22. Hertelé, S.; De Waele, W.; Verstraete, M.; Denys, R.; O'Dowd, N. *J*-integral analysis of heterogeneous mismatched girth welds in clamped single-edge notched tension specimens. *Int. J. Press. Vessel. Pip.* **2014**, *119*, 95–107. [CrossRef]
23. Hertelé, S.; O'Dowd, N.; Van Minnebruggen, K.; Verstraete, M.; De Waele, W. Fracture Mechanics Analysis of Heterogeneous Welds: Validation of a Weld Homogenisation Approach. *Procedia Mater. Sci.* **2014**, *3*, 1322–1329. [CrossRef]
24. Souza, R.F.; Ruggieri, C.; Zhang, Z. A framework for fracture assessments of dissimilar girth welds in offshore pipelines under bending. *Eng. Fract. Mech.* **2016**, *163*, 66–88. [CrossRef]
25. Systemes, D. *Abaqus Analysis User Manual*; Dassault Systemes: Vélizy-Villacoublay, France, 2018.
26. Anderson, T.L. *Fracture Mechanics—Fundamental and Applications*, 3rd ed.; CRC Press LLC, Taylor & Francis Group: Boca Raton, FL, USA, 2005.
27. *ASTM E399-12e3*; Standard Test Method for Linear-Elastic Plane-Strain Fracture Toughness KIc of Metallic Materials. ASTM International: West Conshohocken, PA, USA, 2013.
28. Wu, S.X.; Mai, Y.W.; Cotterell, B. Plastic η-factor (ηp). *Int. J. Fract.* **1990**, *45*, 1–18. [CrossRef]
29. Nevalainen, M.; Dodds, R.H. Numerical investigation of 3-D constraint effects on brittle fracture in SE(B) and C(T) specimens. *Int. J. Fract.* **1995**, *74*, 131–161. [CrossRef]
30. Kim, Y.J.; Kim, J.S.; Cho, S.M.; Kim, Y.J. 3-D constraint effects on *J* testing and crack tip constraint in M(T), SE(B), SE(T) and C(T) specimens: Numerical study. *Eng. Fract. Mech.* **2004**, *71*, 1203–1218. [CrossRef]

Article

The Effect of Material Heterogeneity and Temperature on Impact Toughness and Fracture Resistance of SA-387 Gr. 91 Welded Joints

Milivoje Jovanović [1], Ivica Čamagić [1], Simon Sedmak [2], Aleksandar Sedmak [3,*] and Zijah Burzić [4]

[1] Department for Mechanics, Faculty of Technical Sciences, University of Priština Temporarily Settled in Kosovska Mitrovica, 38220 Kosovska Mitrovica, Serbia; milivoje.s.jovanovic@gmail.com (M.J.); ivica.camagic@pr.ac.rs (I.Č.)

[2] Innovation Center of the Faculty of Mechanical Engineering, 11000 Belgrade, Serbia; simon.sedmak@yahoo.com

[3] Faculty of Mechanical Engineering, University of Belgrade, 11000 Belgrade, Serbia

[4] Military Technical Institute, 11000 Belgrade, Serbia; zijah.burzic@vti.vs.rs

* Correspondence: aleksandarsedmak@gmail.com

Abstract: This paper presents the analysis of the behavior of welded joints made of 9–12% Cr-Mo steel SA-387 Gr. 91. The successful application of this steel depends not only on the base metal's (BM) properties but even more on heat-affected-zone (HAZ) and weld metal (WM), both at room and at operating temperature. Impact testing of specimens with a notch in BM, HAZ, and WM was performed on a Charpy instrumented pendulum to enable the separation of the total energy in crack-initiation and crack-propagation energy. Fracture toughness was also determined for all three zones, applying standard procedure at both temperatures. Results are analyzed to obtain a deep insight into steel SA 387 Gr. 91's crack resistance properties at room and operating temperatures. Results are also compared with results obtained previously for A-387 Gr. B to assess the effect of an increased content of Chromium.

Keywords: welded joint; crack-initiation energy; crack-propagation energy; fracture toughness

Citation: Jovanović, M.; Čamagić, I.; Sedmak, S.; Sedmak, A.; Burzić, Z. The Effect of Material Heterogeneity and Temperature on Impact Toughness and Fracture Resistance of SA-387 Gr. 91 Welded Joints. *Materials* **2022**, *15*, 1854. https://doi.org/10.3390/ma15051854

Academic Editors: Tomasz Trzepieciński, Nenad Gubeljak and Dražan Kozak

Received: 17 November 2021
Accepted: 27 January 2022
Published: 2 March 2022

Publisher's Note: MDPI stays neutral with regard to jurisdictional claims in published maps and institutional affiliations.

1. Introduction

The Cr-Mo steel SA-387 Gr. 91 belongs to a group of heat-and creep-resistant 9–12% Cr steels. They are introduced into practice to replace 2.25% Cr-Mo steel for operating temperatures above 565 °C, with the maximum service temperature of Gr. 91 equaling circa 600 °C [1].

The Cr-Mo steel SA-387 Gr. 91 has exceptional mechanical properties, including crack initiation and propagation resistance, making it an excellent choice for pressure vessels operating at elevated temperatures. It is a simple matter to compare SA-387 Gr. 91, in its role as a base metal, with more conventional steels, such as SA-387 Gr. B, and to find out that conventional design methods will lead to significant reductions in pressure vessel thickness and costs in general [1]. However, having in mind the importance and complexity of welded joints, it is of utmost importance to obtain a deep insight into the behavior of all zones (base metal, weld metal, heat-affected zone) to ensure the safe application and exploitation of Cr-Mo steel. Knowing that welded joint crack sensitivity increases, the more complex the composition and structure of a steel becomes, it is reasonable to assume that a complete overview of SA 387 Gr. 91 application must include a detailed analysis of its crack resistance in all welded joint zones. This should include at least Charpy toughness and fracture toughness testing, both at room and elevated temperatures, up to 575 °C, which is the aim of this research.

In a limited number of papers published about the effect of material heterogeneity and temperature on steel SA-387 Gr. 91's behavior, most of the focus was on strength and creep properties in relation to their microstructure, especially in a HAZ [2–11]. Detailed study of the simulated heat-affected zone of creep-resistant 9–12% advanced chromium steel is presented in [2–5], with an emphasis on the microscopic analysis of the influence of multiple thermal cycles on simulated HAZ toughness [3], the relationship between microstructure and mechanical properties [4], and the significance of cracks [5]. The fracture properties of different microstructural regions of the heat-affected zone (HAZ) of modified 9Cr-1Mo steel (tempered base metal, inter-critical, fine grained, coarse grained with and without δ-ferrite) have been also studied using the Charpy impact test in [6]. A simulation technique is used to reproduce HAZ microstructures. The results indicated the lowest toughness in coarse-grained regions of the HAZ [6].

One of the most important problems with SA-387 Gr. 91 is its narrow heat-affected zone with heterogeneous structures in the base metal, generated due to non-equilibrium phase transformations during the arc-welding processes [7–9]. This heterogeneous heat-affected zone has been reported to cause the short-term creep failures of welded components, such as the infamous Type IV cracking [6]. It was shown in [7–9] that the Post-Weld Heat Treatment (PWHT) plays a significant role in improving weldments' toughness and maximizing their creep lifetime. A similar study was made on somewhat different steel, Gr 92 and G92N, with a focus on the effect of Boron and Nitrogen, as presented in [10].

The creep-crack growth behavior of a P92 steel-welded joint was analyzed in [11] with the crack tip located at different distinct zones of welded joint. Tested results revealed that even in thin thickness specimens, fine-grained heat-affected zone specimens exhibited a fast creep-crack growth rate compared with other micro-zone specimens due to a low creep crack resistance and a high multi-stress state.

The approach used in this paper had already been applied in the case of structural, low-alloyed HSLA steel and its welded joint constituents [12], as well as in a series of papers presenting research on a similar steel, A 387 Gr. B, with 1% Cr, which is also used for elevated temperature [13–18]. In general, all relevant mechanical properties of steel A 387 Gr. B, such as tensile properties, Charpy impact toughness, fracture toughness, and Paris law coefficients, have been presented in series of papers [13–19], including the effects of time and temperature. More concretely, the influence of temperature and exploitation period (time) on the behavior of a welded joint subjected to impact loading was analyzed in [13], while the same effects on plane strain fracture toughness in a welded joint were analyzed in [14], indicating very good crack resistance properties of all three regions, with small differences between them. One interesting approach to measuring the relationship between the impact and fracture toughness of A-387 Gr. B welded joint was presented in [15], where separated energies as obtained using Charpy instrumented pendulum were compared with fracture toughness. The effect of temperature and exploitation time on tensile properties and plane strain fracture toughness in a welded joint was analyzed in [16,17], also indicating the good properties of all three zones, i.e., BM, WM, and HAZ. The influence of temperature and exploitation period on fatigue-crack growth parameters in different regions of welded joints was analyzed in [18], indicating the highest crack-growth-rate values in HAZ and the lowest in BM. Therefore, the lowest fatigue-crack resistance of steel A-387 Gr. B is in HAZ. Moreover, higher temperatures and longer exploitation periods increase crack growth rates and decrease fatigue thresholds for both new and exploited materials in all regions of welded joints (BM, WM, HAZ). These effects occur as the result of microstructural changes, such as carbide formation and growth at grain boundaries and inside grains [18]. Finally, a recently published paper [19], dealt with the crack resistance of SA 387 Gr. 91 welded joints under static and impact load, presenting preliminary results for Charpy impact toughness and fracture toughness, indicating good resistance to crack growth for all three welded joint regions, but with significant differences between them. As expected, crack resistance in HAZ is reduced compared to BM, and, somewhat less expected, WM is significantly more sensitive to cracking but still performs at a satisfying level. In this paper, more results for Charpy impact toughness and fracture toughness of SA 387 Gr. 91 welded joints and more detailed analysis will be presented, with a focus on material heterogeneity and temperature effects, as already briefly outlined in [19], but also with a focus on the Chromium effect, which was not previously analyzed in this way. Therefore, the results presented here will not be only analyzed on their own, but also in comparison with corresponding results for A-387 Gr. B to obtain a better insight into the effect of a significantly increased content level of Chromium. In any case, the focus in this research is on the effect of weldment heterogeneity (i.e., the different properties of BM, WM, and HAZ) and Cr content on the crack resistance of a welded joint.

2. Methods

Steel SA-387 Gr. 91 is designed to have the minimum yield stress of 450 MPa and minimum impact energy of 41 J at room temperature, with the idea that is will be able to work at elevated temperatures with sufficient strength and toughness [1,19]. For this research, steel SA-387 Gr. 91, thickness 15 mm, produced in "Steelwork ACRONI" Jesenice Slovenia, was used, with the chemical composition shown in Table 1.

Welding was performed in 4 root and 10 filler passes, using a Gas Tungsten Arc Welding (GTAW) with BOEHLER C9 MV-IG Ø2.4 mm filler metal to ensure high quality of root passes 1–4 and a Shielded Metal Arc Welding (SMAW) with BOEHLER FOX C9 MV electrode, diameters 2.5 (passes 5–9) and 3.25 mm (passes 10–14), as shown schematically in Figure 1. The chemical compositions and mechanical properties of filler metals are shown in Table 2. Welding parameters and linear energies (as shown in Table 3) were chosen carefully to adjust cooling speed and optimize welded-joint microstructure, with thermal efficiency coefficients taken as 0.6 (GTAW) and 0.8 (SMAW).

Pre-heating was performed at 250 °C, while the inter-pass temperature was 200–300 °C. Post-Weld Heat Treatment (PWHT) was applied, consisting of tempering at 250 °C, followed by heating up to 750 °C (rate 100–150 °C/h), holding at 750 for 2 h, and cooling down to 400 °C (rate 150 °C/h), with final cooling at the still air.

Table 1. Base metal chemical composition, steel SA-387 Gr. 91.

Chemical Composition, Weight %										
C	Si	Mn	P	S	Cr	Mo	Ni	V	Nb	Cu
0.129	0.277	0.443	0.001	0.001	8.25	0.874	0.01	0.198	0.056	0.068

Table 2. Filler metal chemical composition (%).

Filler Metal	C	Si	Mn	P	S	Cr	Mo	Ni	V	Nb	Cu
C9 MV IG Ø2.4 mm	0.11	0.23	0.5	0.006	0.003	9.0	0.93	0.5	0.19	0.07	0.0
FOX C9 MV Ø2.5 mm	0.09	0.19	0.55	0.01	0.006	8.5	1.0	0.5	0.19	0.04	0.1
FOX C9 MV Ø3.25 mm	0.11	0.26	0.66	0.008	0.005	8.5	0.94	0.5	0.20	0.06	0.1

Table 3. Welding parameters and linear energies.

Pass	Voltage V	Current A	Welding Speed mm/s	Linear Energy kJ/mm
1	12.2	172	0.3	4.2
2	12.2	172	0.6	2.1
3–4	12.2	172	0.9	1.4
5–9	25.4	126	3.0	0.85
10–14	25.4	126	2.6	0.98

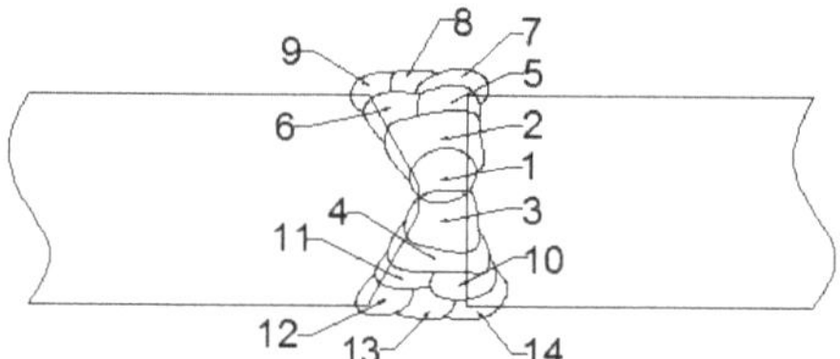

Figure 1. Welded plate (thickness 15 mm) cross-section, welding passes: root 1–4, filler 5–14.

2.1. Impact Testing on Charpy Instrumented Pendulum

The testing procedure was applied according to SRPS EN ISO 9016:2013 [20], including specimen shape and size, as well as notch V-2 position, Figure 2. Testing was performed on an instrumented Charpy pendulum SCHENCK TREBELL 150/300 J at room temperature, 20 °C, and an elevated temperature, 575 °C. The higher temperature was chosen as it is a common service temperature for this steel, whereas room temperature was used as a reference, so that the effect of high temperature on the steel could be evaluated. Three specimens were extracted from each characteristic zone with the crack tip positioned in the BM, WM, and HAZ. In the case of the BM, the specimens were taken from a location far from the weld metal, as a common practice to avoid welding heat effects. In the case of the HAZ, the specimen tip was located in the HAZ, as close to the WM as possible (as shown in Figure 2), since the CGHAZ was shown to have the lowest toughness in HAZ [6]. In the case of the WM specimen, the tip was located close to the center line.

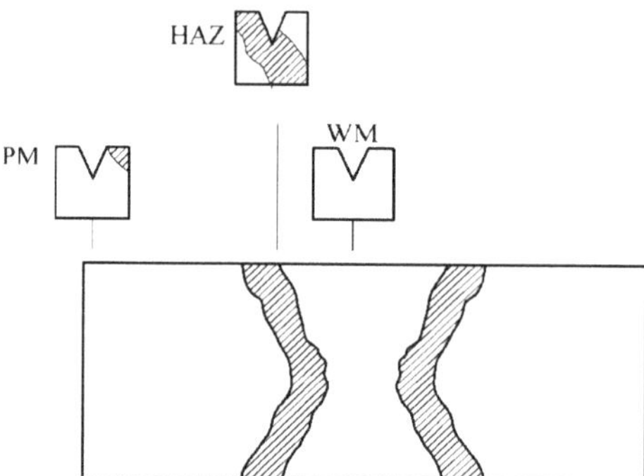

Figure 2. Welded joint Charpy specimen cutting scheme.

Since the tests were performed using an instrumented Charpy pendulum, it was possible to separate crack initiation and propagation energies, and to evaluate the effect of the notch location on the impact properties and plasticity. In this way, it was possible to determine the energy required for initiating a crack and the energy required for its propagation, enabling better understanding of the crack resistance of tested material, as explained in more details in [21], including different methods to separate these two energies. In this research, separation was performed according to the force maximum value, so that the area to the left represents the energy for crack initiation, A_i, the area to the right the energy for crack propagation, A_p, Figure 3.

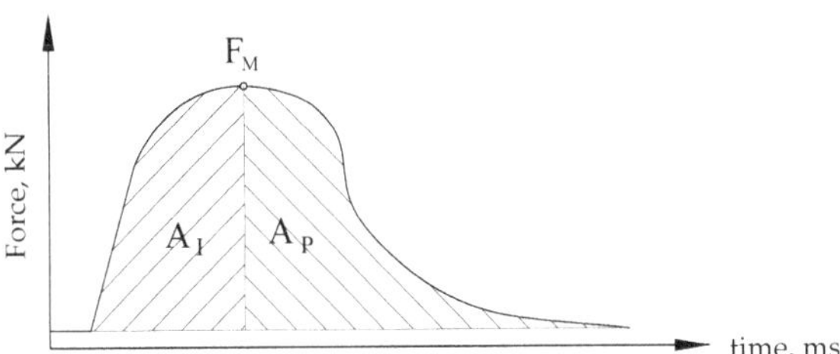

Figure 3. Separation of energies for crack initiation and propagation.

2.2. Fracture Toughness, K_{Ic}, Testing

Three-point single-edge bending (SEB) specimens were used for fracture toughness, K_{Ic}, measurement at room temperature, whereas modified CT specimens were used at elevated temperature, 575 °C (as shown in Figure 4). This modification was needed due to the shape of the chamber used for testing at 575 °C and had no effect on fracture toughness values, since the stress–strain state at the crack tip was not affected. Fracture toughness, K_{Ic}, was determined via critical J integral, J_{Ic}, applying rules of elastic–plastic fracture mechanics (EPFM) [22]:

$$K_{Ic} = \sqrt{\frac{J_{Ic} \cdot E}{1 - v^2}},\tag{1}$$

where E is the Elasticity modulus, and is v the Poisson ratio. Standard procedure is defined in the ASTM 1820 standard [23] along with the specific aspects for a welded joint testing set out in [24]. A crack was produced on the HF testing machine, and its length was measured after the experiment, as defined in [23].

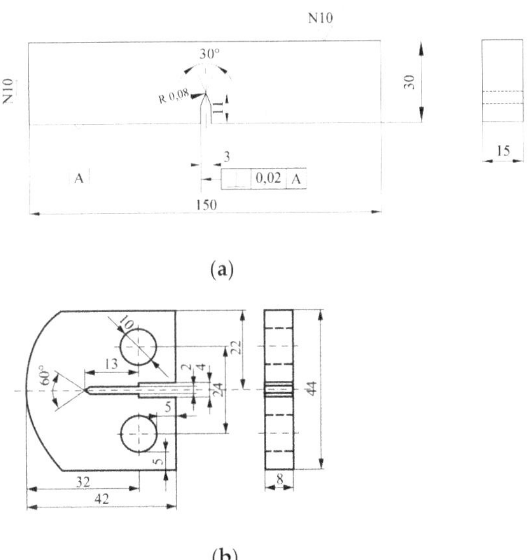

(**a**)

(**b**)

Figure 4. Specimens for K_{Ic} testing. (**a**) SENB, (**b**) CT.

3. Results

3.1. Impact Testing

Results of the impact tests are provided for BM, WM, and HAZ in Tables 4–6, respectively. Characteristic examples of F-t diagrams are shown in Figures 5–7 for BM, WM, and HAZ, respectively. As one can see from the results presented in Tables 4–6, impact toughness at room temperature is the highest in BM, closely followed by HAZ. High resistance to cracking in HAZ is even more pronounced when energy components are considered, since it has the highest resistance to crack initiation. In any case, one should keep in mind that all zones in SA 387 Gr. 91 have a relatively high impact energy, both for crack initiation and propagation, making their welded joints resistant to cracking. At this point, one should notice that such result actually leads to the conclusion that the welding procedure specification for SA 387 Gr. 91 is well defined, and welding itself is well performed.

The results of the impact testing are in good agreement with the presented microstructures and hardness values, since the highest impact energy (BM) corresponds with the lowest hardness, and the lowest impact energy (WM) corresponds with the highest hardness.

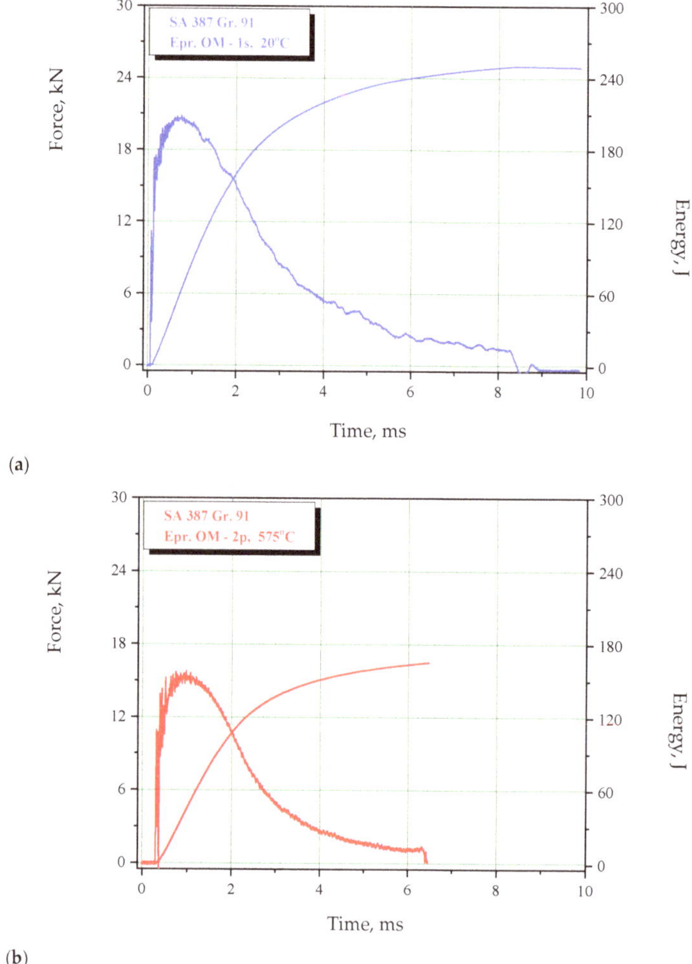

Figure 5. Charpy instrumented pendulum diagrams for BM, SA-387 Gr. 91 at (**a**) 20 °C, (**b**) 575 °C.

Table 4. Results of Charpy testing—SA 387 Gr. 91 BM.

Specimen	Testing Temperature, °C	Impact Total Energy, A_T, J	Crack-Initiation Energy, A_I, J	Crack-Growth Energy, A_P, J
BM-1A		251	58	193
BM-2A	20	268	60	208
BM-3A		275	58	217
average		265	59	206
BM-4A		159	41	118
BM-5A	575	166	43	123
BM-6A		155	41	114
average		160	42	118

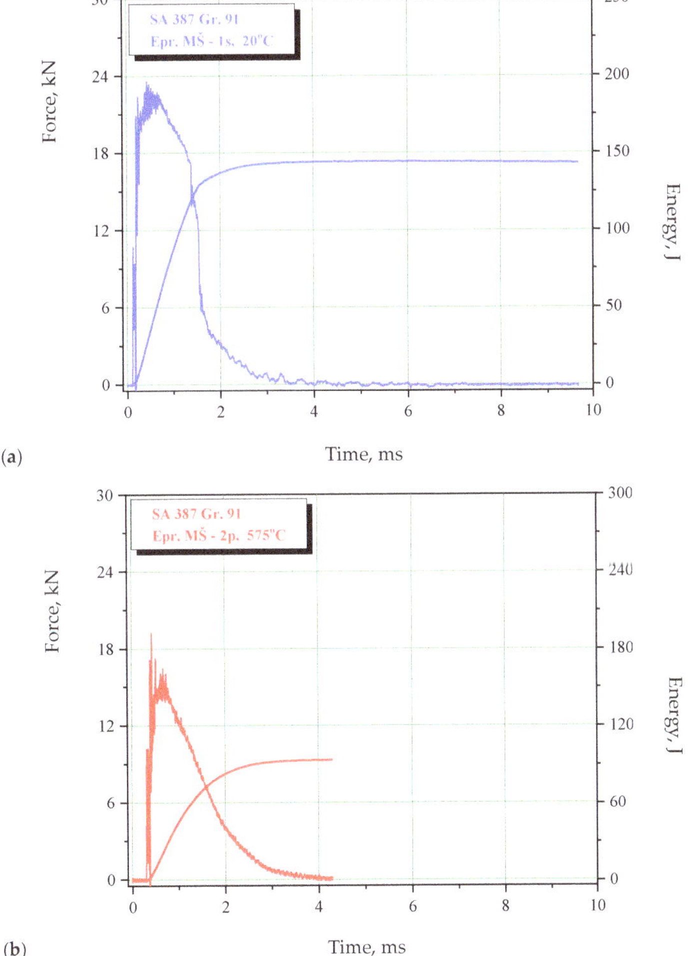

(**a**)

(**b**)

Figure 6. Charpy instrumented pendulum diagrams for WM, SA-387 Gr. 91 at (**a**) 20 °C, (**b**) 575 °C.

Table 5. Results of Charpy testing—SA 387 Gr. 91 WM.

Specimen Mark	Testing Temperature, °C	Impact Total Energy, AT, J	Crack-Initiation Energy, AI, J	Crack-Growth Energy, AP, J
WM-1A		144	52	92
WM-2A	20	168	55	113
WM-3A		156	52	104
average		156	53	103
WM-4A		92	28	64
WM-5A	575	94	28	66
WM-6A		104	29	75
average		97	28	69

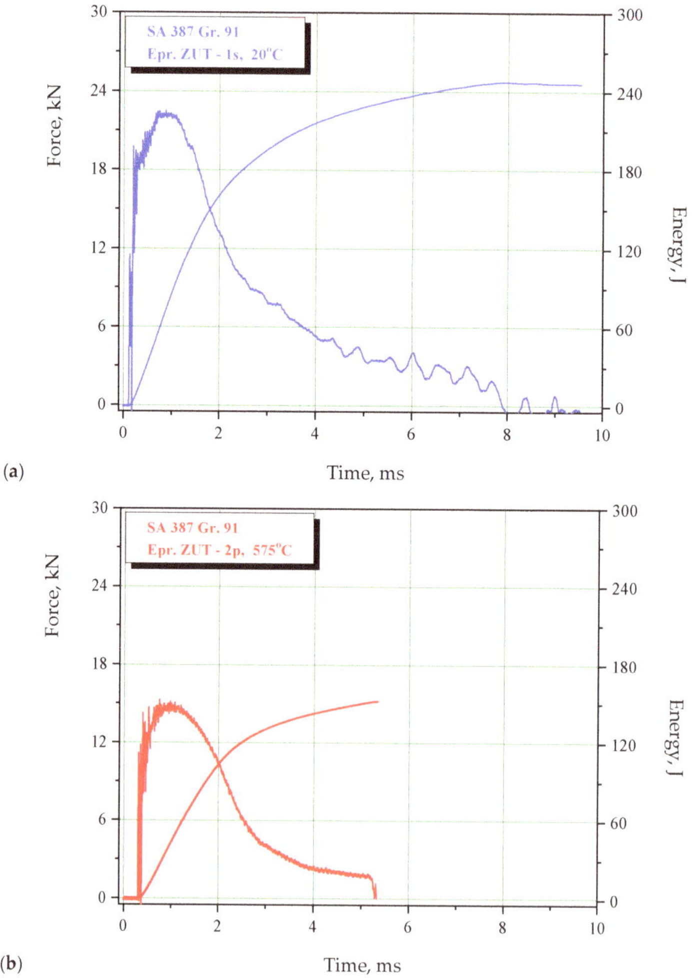

Figure 7. Charpy instrumented pendulum diagrams for HAZ, SA-387 Gr. 91 at (**a**) 20 °C, (**b**) 575 °C.

Table 6. Results of Charpy testing—SA 387 Gr. 91 HAZ.

Specimen Mark	Testing Temperature, °C	Impact Total Energy, A_T, J	Crack-Initiation Energy, A_I, J	Crack-Growth Energy, A_P, J
HAZ-1A		248	70	178
HAZ-2A	20	246	69	177
HAZ-3A		248	70	178
average		248	70	178
HAZ-4A		147	39	108
HAZ-5A	575	153	42	111
HAZ-6A		138	40	98
average		146	40	106

Testing at the operating temperature indicates similar behavior, since the reduction of energies is similar: BM 29–43%, WM 26–47%, and HAZ 9–43%. Therefore, energy values at the operating temperature, compared with the room temperature are as follows: BM 57–71%, WM 53–74%, HAZ 57–91%, which are still relatively high. The lowest value is energy for crack initiation, $A_I = 28$ J, which was recorded in WM and is still reasonable from a practical point of view.

The effect of different zones in a welded joint on crack initiation and propagation is also visible in Figures 8–10, where fractographies of BM, WM, and HAZ are shown, respectively, for both testing temperatures. It is clear that only Figures 8a, 9a and 10a, which present the crack initiation process at 575 °C, do not show completely ductile fracture surfaces, which is in agreement with the lower energies recorded for crack initiation in these specimens (42, 28, and 40 J, respectively). However, they do not represent brittle fractures either; all fractographies indicate sufficient toughness values and high resistance to crack initiation and propagation. Moreover, one should notice relatively small differences in crack initiation and propagation energies between the different zones in a welded joint made of SA387 Gr. 91, which is also proved by the presented fractographies.

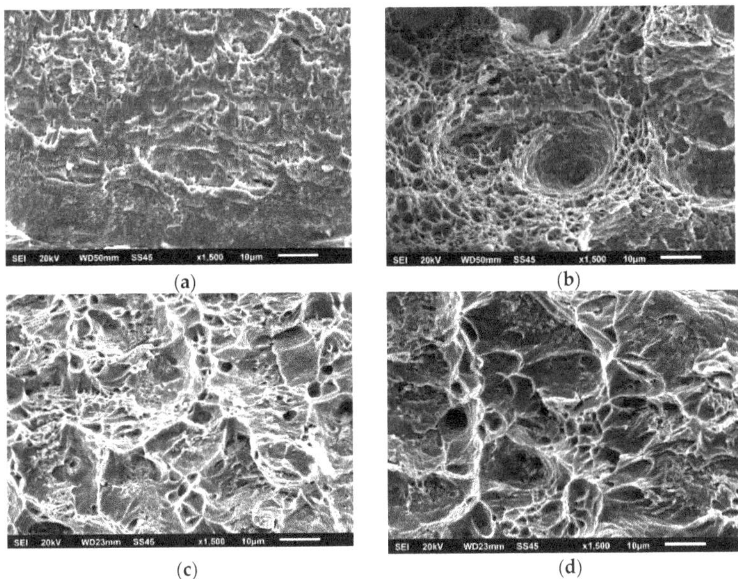

Figure 8. SA-387 Gr. 91 BM fractography upon (**a**) crack initiation, 575 °C; (**b**) crack propagation, 575 °C; (**c**) crack initiation, 20 °C; and (**d**) crack propagation, 20 °C.

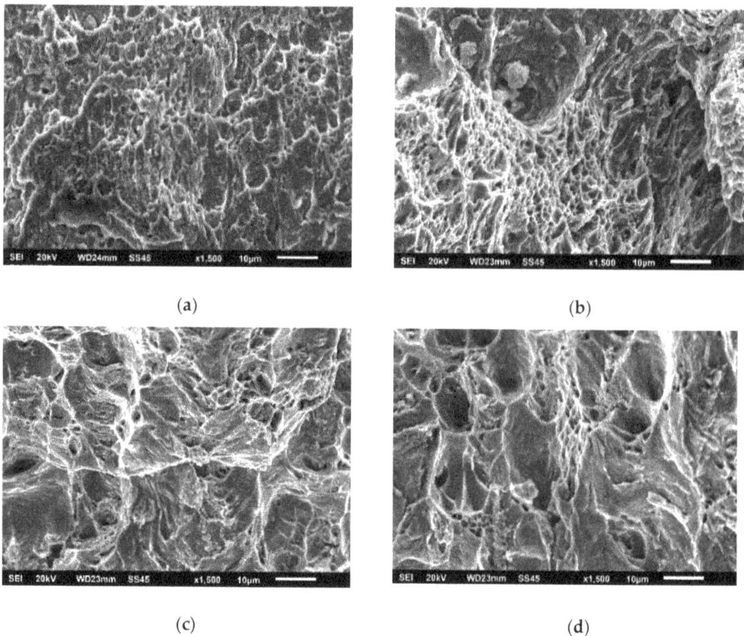

(a)

(b)

(c)

(d)

Figure 9. SA-387 Gr. 91 WM fractography upon (**a**) crack initiation, 575 °C; (**b**) crack propagation, 575 °C; (**c**) crack initiation, 20 °C; and (**d**) crack propagation, 20 °C.

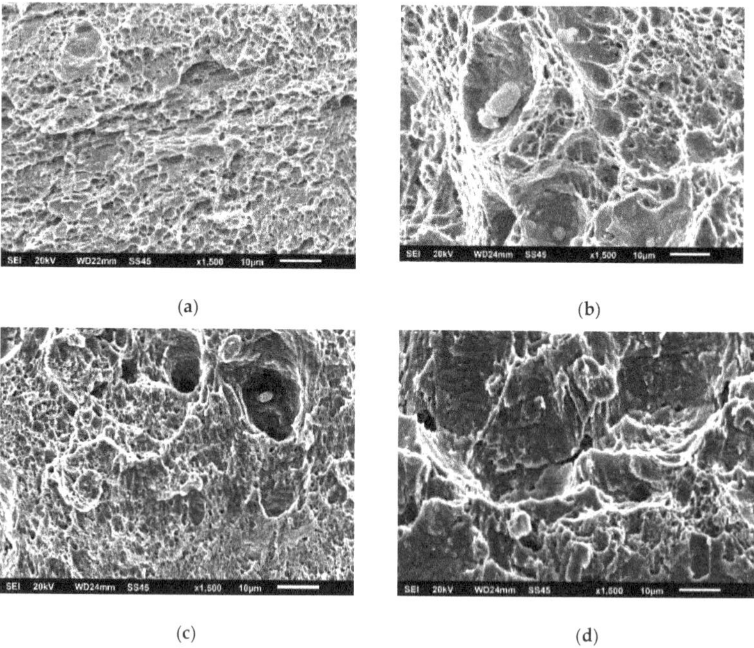

(a)

(b)

(c)

(d)

Figure 10. SA-387 Gr. 91 HAZ fractography upon (**a**) crack initiation, 575 °C; (**b**) crack propagation, 575 °C; (**c**) crack initiation, 20 °C; and (**d**) crack propagation, 20 °C.

The results of the impact testing for A387 Gr. B, obtained at room temperature, are presented in Tables 7–9 for BM, WM, and HAZ, respectively. The distribution of energies, both total and separated, is similar as for 9% Cr steel, the highest values are in BM, but are followed closely by both HAZ and WM, in this case. These results also lead to the conclusion that the welding procedure specification for A 387 Gr. B is well defined, and the welding itself is well performed.

The reduction of energy at operating temperature is smaller for steel with 1% Cr than for steel with 9% Cr, with similar distribution of energy: BM 18–38%, WM 8–30%, HAZ 20–29%. The levels of energy in relation to the room temperature are: BM 62–82%, WM 70–92%, HAZ 71–80%. The lowest individual value for initial energy, $A_I = 38$ J, is recorded in BM and is still satisfactory. Nevertheless, one should not forget that the operating temperature for 1% Cr steel is 540 °C, i.e., lower than that for 9% Cr (575 °C), so the reduction of energies was expected, not only because of the simpler microstructure (less Cr).

Table 7. Results of impact testing for A387 Gr. B—BM [13].

Specimen Mark	Testing Temperature, °C	Impact Total Energy, A_T, J	Crack-Initiation Energy, A_I, J	Crack-Propagation Energy, A_P, J
BM-1-1n		204	47	157
BM-1-2n	20	212	49	163
BM-1-3n		214	49	165
average		210	48	162
BM-2-1n		137	38	99
BM-2-2n	540	139	40	99
BM-2-3n		145	41	104
average		141	40	101

Table 8. Results of impact testing for A387 Gr. B/WM [13].

Specimen Mark	Testing Temperature, °C	Impact Total Energy, A_T, J	Crack-Initiation Energy, A_I, J	Crack-Propagation Energy, A_P, J
WM-1-1		193	56	137
WM-1-2	20	190	60	130
WM-1-3		183	60	123
average		189	59	130
WM-2-1		139	40	99
WM-2-2	540	133	39	94
WM-2-3		134	39	95
average		135	39	96

Table 9. Results of impact testing for A387 Gr. B—HAZ [13].

Specimen Mark	Testing Temperature, °C	Impact Total Energy, A_T, J	Crack-Initiation Energy, A_I, J	Crack-Propagation Energy, A_P, J
HAZ-1-1e		186	47	139
HAZ-1-2e	20	187	45	142
HAZ-1-3e		183	47	136
average		185	46	139
HAZ-2-1e		143	46	97
HAZ-2-2e	540	131	43	88
HAZ-2-3e		129	42	87
average		134	44	90

3.2. Fracture Toughness Testing

Fracture toughness values are obtained via J_{Ic}, as explained in Section 2.2, using J-R curves as shown in Figures 11–13 for characteristic examples of BM, WM, and HAZ testing, respectively. Calculated K_{Ic} values for SA 387 Gr. 91 steel are provided in Tables 10–12 for BM, WM, and HAZ, respectively, clearly indicating that the K_{Ic} values are satisfactory, with the highest values in BM (175.0 and 124.4 J for the room and operating temperature, respectively), the lowest in WM (125.7 and 91.1 J), and in-between in HAZ (146.4 and 111.9 J). The effect of heterogeneity and temperature is similar, as in the case of impact toughness, with HAZ being somewhat more sensitive to cracking, and with a slightly smaller reduction of K_{Ic} values (the ratio between 20 and 575 °C values is circa 1.3 compared to circa 1.6 for impact toughness) with increased temperature.

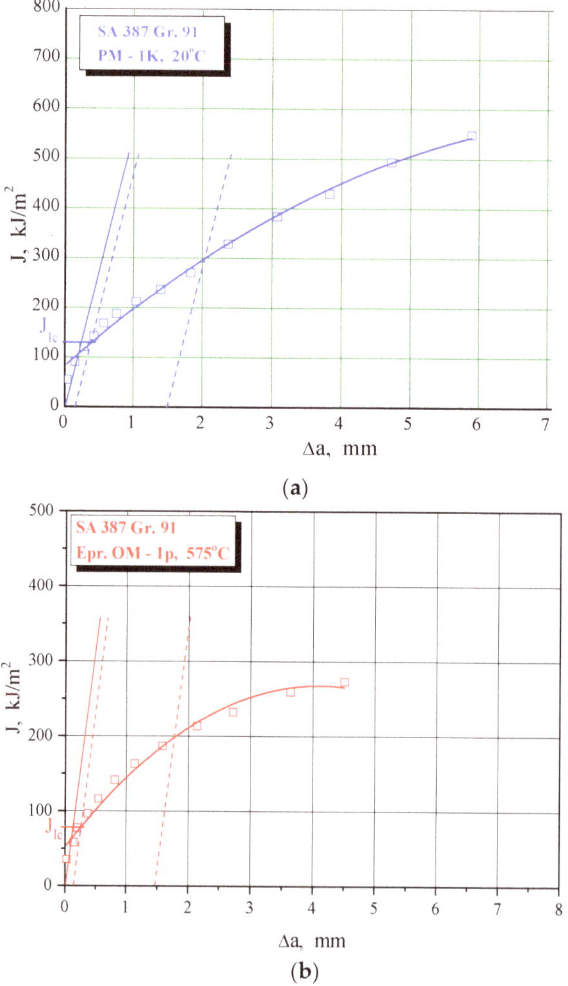

Figure 11. The J-R curve and J_{Ic} evaluation for SA 387 Gr. 9—BM at (**a**) 20 °C, (**b**) 575 °C.

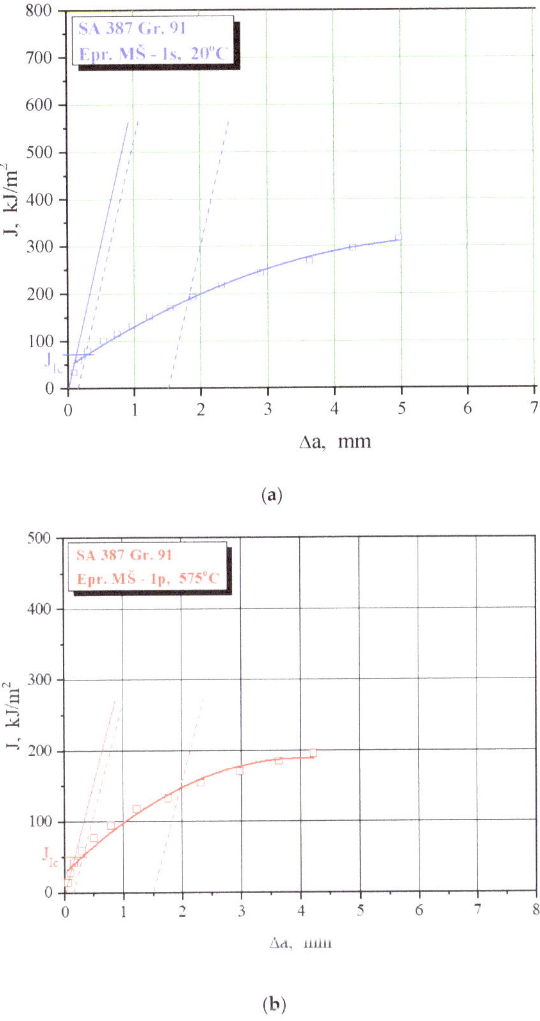

Figure 12. The J-R curve and J_{Ic} evaluation for SA 387 Gr. 91—WM at (**a**) 20 °C, (**b**) 575 °C.

Table 10. K_{Ic} values obtained via J_{Ic} for SA 387 Gr. 91—BM.

Specimen Mark	Testing Temperature, °C	Critical J-Integral, J_{Ic}, kJ/m²	Critical Stress Intensity Factor, K_{Ic}, MPa·m$^{1/2}$
BM-1K		131.1	173.9
BM-2K	20	144.2	182.4
BM-3K		124.0	169.2
average			175.0
BM-4K		78.5	122.9
BM-5K	575	80.9	124.7
BM-6K		81.9	125.5
average			124.4

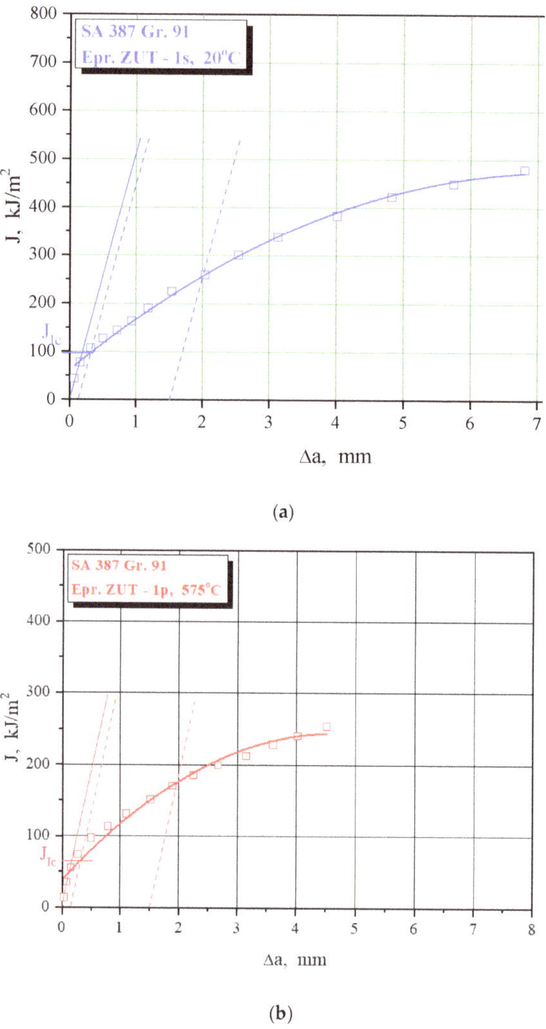

(a)

(b)

Figure 13. The J-R curve and J_{Ic} evaluation for SA 387 Gr. 91—HAZ at (**a**) 20 °C, (**b**) 575 °C.

Table 11. K_{Ic} values obtained via J_{Ic} for SA 387 Gr. 91—WM.

Specimen Mark	Testing Temperature, °C	Critical J-Integral, J_{Ic}, kJ/m²	Critical Stress Intensity Factor, K_{Ic}, MPa·m$^{1/2}$
WM-1K		71.6	128.5
WM-2K	20	64.8	122.3
WM-3K		69.2	126.4
average			125.7
WM-4K		51.2	99.2
WM-5K	575	40.1	87.8
WM-6K		38.6	86.2
average			91.1

Table 12. K_{Ic} values obtained via J_{Ic} for SA 387 Gr. 91—HAZ.

Specimen Mark	Testing Temperature, °C	Critical J-Integral, J_{Ic}, kJ/m^2	Critical Stress Intensity Factor, K_{Ic}, MPa·m$^{1/2}$
HAZ-1K		97.6	150.1
HAZ-2K	20	88.9	143.2
HAZ-3K		92.1	145.8
average			146.4
HAZ-4K		65.3	112.1
HAZ-5K	575	61.6	108.8
HAZ-6K		68.5	114.8
average			111.9

As in the case of impact toughness, one should notice relatively small differences in K_{Ic} values between the different zones in welded joints made of SA387 Gr. 91, proved also by the presented fractographies (as shown in Figures 14–16), indicating sufficiently ductile material. Even in the case of WM at 575 °C (as shown in Figure 15b), which appears to be a brittle fracture, it was found that this is actually a 'local brittle zone' (LBZ), not uncommon for WM, especially if made of alloyed steel. The same fractography, but with a magnification of 200×, is shown in Figure 17, also indicating some typical features of a ductile fracture.

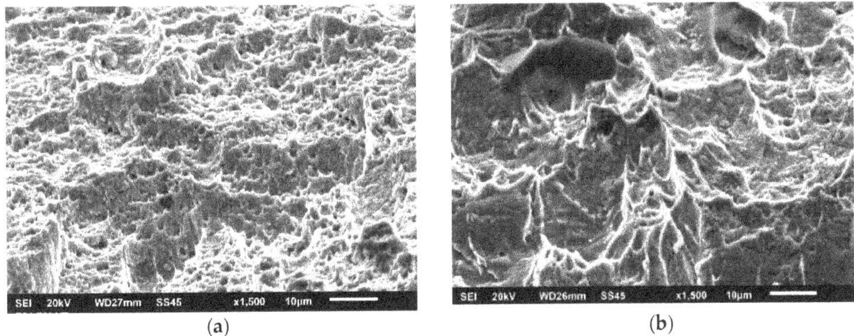

(a) (b)

Figure 14. SA-387 Gr. 91 BM fractography. (a) SENB specimen 20 °C, (b) CT specimen 575 °C, 1500×.

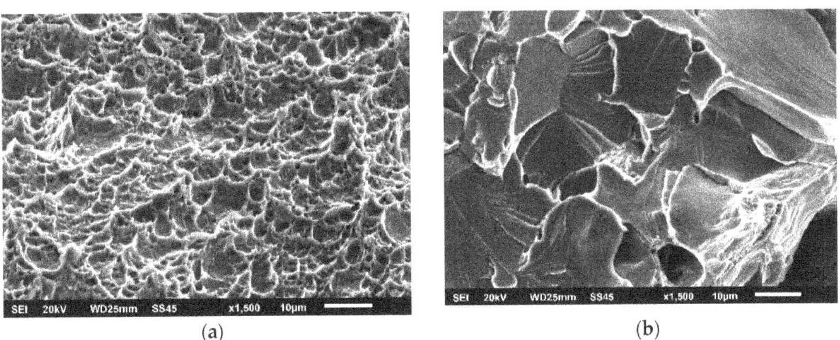

(a) (b)

Figure 15. SA-387 Gr. 91 WM fractography. (a) SENB specimen 20 °C, (b) CT specimen 575 °C, 1500×.

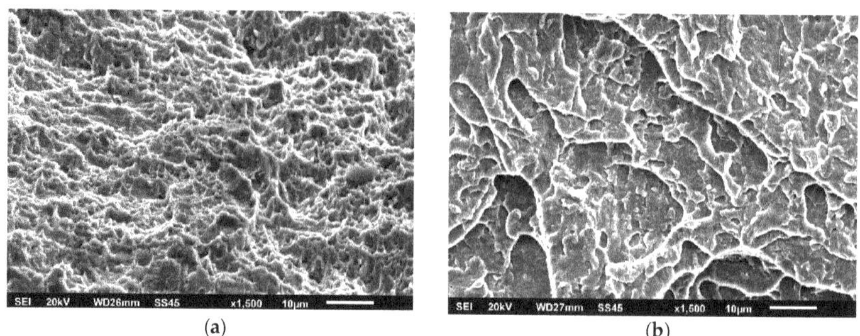

(a)   (b)

Figure 16. SA-387 Gr. 91 HAZ fractography. (**a**) SENB specimen 20 °C, (**b**) CT specimen 575 °C, 1500×.

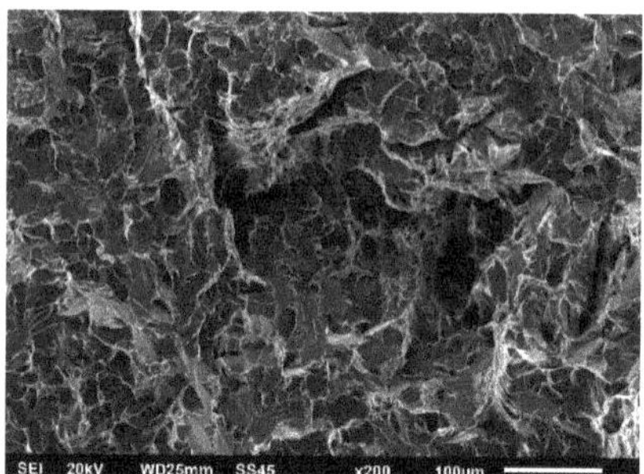

Figure 17. SA-387 Gr. 91 WM fractography, CT specimen 575 °C, 200×.

Values for K_{Ic} in the case of A387 Gr. B steel are provided for comparison in Tables 13–15 for BM, WM, and HAZ, respectively. In this case, the results for fracture toughness are different from those for impact toughness in two important aspects: the first is that 1% Cr steel has lower values than 9% Cr, and the second is that WM is now the region with the highest crack resistance, whereas HAZ is the weakest link. However, the differences are very small, since the average values for WM are circa 10% higher, and for HAZ circa 10% lower, than BM average values. Therefore, the effect of heterogeneity is less pronounced than in the case of SA 387 Gr. 91 steel, whereas the effect of temperature is slightly stronger (the ratio between 20 °C and 575 °C values is circa 1.4, which is almost the same as in the case of SA 387 Gr. 91).

Table 13. Values of K_{Ic} via J_{Ic} for A387 Gr. B—BM.

Specimen Mark	Testing Temperature, °C	Critical J-Integral, J_{Ic}, kJ/m^2	Critical Stress Intensity Factor, K_{Ic}, MPa·m$^{1/2}$
BM-1-1n		60.1	117.8
BM-1-2n	20	63.9	121.4
BM-1-3n		58.6	116.3
average			118.5
BM-2-1n		43.2	87.2
BM-2-2n	540	44.7	88.7
BM-2-3n		45.3	89.2
average			85.7

Table 14. Values of K_{Ic} via J_{Ic} for A387 Gr. B—WM.

Specimen Mark	Testing Temperature, °C	Critical J-Integral, J_{Ic}, kJ/m^2	Critical Stress Intensity Factor, K_{Ic}, MPa·m$^{1/2}$
WM-1-1		72.8	129.6
WM-1-2	20	74.3	130.9
WM-1-3		71.1	128.1
average			129.5
WM-2-1		50.2	93.9
WM-2-2	540	52.6	96.2
WM-2-3		48.4	92.2
average			94.1

Table 15. Values of K_{Ic} via J_{Ic} for A387 Gr. B—HAZ.

Specimen Mark	Testing Temperature, °C	Critical J-Integral, J_{Ic}, kJ/m^2	Critical Stress Intensity Factor, K_{Ic}, MPa·m$^{1/2}$
HAZ-1-1n		53.6	111.2
HAZ-1-2n	20	51.7	109.2
HAZ-1-3n		49.8	107.2
average			109.2
HAZ-2-1n		33.6	76.9
HAZ-2-2n	540	34.2	77.5
HAZ-2-3n		36.1	79.7
average			78.0

4. Conclusions

Based on the results presented in this paper, one can conclude the following:

- Both steels, SA-387 Gr. 91 and A-387 Gr. B, as well as their welded joints, have high resistance to cracking, both for static and impact loading. This conclusion also holds for SA 387 Gr. 91 WM, even though its resistance to cracking is lower than BM and HAZ, but well above 41 J, which is the minimum value for the BM.
- The effect of material heterogeneity on impact toughness is more heavily expressed for SA-387 Gr. 91 than for A-387 Gr. B, since the WM in the former case has lower values of crack initiation and growth energies, whereas these values are balanced in the latter case. A reduction of impact toughness in the case of SA-387 Gr. 91 steel is mostly due to crack-growth energy, which is significantly smaller than for SA-387 Gr. 91 BM and HAZ, but still at a satisfying level.
- The effect of temperature on impact toughness is similar, but more pronounced, since both energies are lower in all cases, approximately 1/3 less than at room temperature, but still at a satisfying level.
- The effect of material heterogeneity on fracture toughness is similar to its effect on impact toughness, but more expressed for SA-387 Gr. 91 than for A-387 Gr. B, for the same reason as in the case of impact toughness. The effect of temperature on fracture toughness is also similar to its effect on impact toughness. One can say that the behavior of both materials and their welded joints in respect to cracking is practically the same for static and impact loading.

Author Contributions: Conceptualization, M.J. and I.Č.; methodology, Z.B.; validation, A.S. and S.S.; formal analysis, Z.B.; investigation, M.J.; resources, M.J.; data curation, I.Č.; writing—original draft preparation, M.J. and A.S.; writing—review and editing, A.S.; visualization, S.S.; supervision, S.S.; project administration, A.S.; funding acquisition, A.S. All authors have read and agreed to the published version of the manuscript.

Funding: This research was funded by the Ministry of Education, Science and Technological Development, Republic of Serbia, grant number 451-03-9/2021-14/200105, 451-03-9/2021-14/200213.

Conflicts of Interest: The authors declare no conflict of interest.

References

1. Milovic, L. Is Substituting P91 for P22 Justified? Fracture at all Scales. In *Mechanical Engineering*; Pluvinage, G., Milovic, L., Eds.; Springer: New York, NY, USA, 2017; pp. 89–103.
2. Milović, L.; Vuherer, T.; Zrilić, M.; Sedmak, A.; Putić, S. Study of the simulated heat affected zone of creep resistant 9–12% advanced chromium steel. *Mater. Manuf. Process.* **2008**, *23*, 597–602. [CrossRef]
3. Łomozik, M.; Zielińska-Lipiec, A. Microscopic analysis of the influence of multiple thermal cycles on simulated HAZ toughness in P91 steel. *Arch. Met. Mater.* **2008**, *53*, 1025–1034.
4. Milović, L.; Vuherer, T.; Blačić, I.; Vrhovac, M.; Stanković, M. Microstructures and mechanical properties of creep resistant steel for application at elevated temperatures. *Mater. Des.* **2013**, *46*, 660–667. [CrossRef]
5. Milovic, L. Significance of cracks in the heat-affected-zone of steels for elevated temperature application. *Struct. Integr. Life* **2008**, *8*, 55–64.
6. Moitra, A. A toughness study of the weld heat-affected zone of a 9Cr-Mo steel. *Mater. Charact.* **2002**, *48*, 55–61. [CrossRef]
7. Wang, Y. Correlation between intercritical heat-affected zone and type IV creep damage zone in Gr. 91 steel. *Metall. Mats. Trans. A* **2018**, *49*, 1264–1275. [CrossRef]
8. Wang, Y.; Kannan, R.; Li, L. Identification and Characterization of Intercritical Heat Affected Zone in As-welded Grade 91 Weldment. *Metall. Mater. Trans. A* **2016**, *47*, 5680–5684. [CrossRef]
9. Wang, Y.; Kannan, R.; Zhang, L.; Li, L. Microstructural Analysis of As-Welded Heat Affected Zone of Grade 91 Steel Heavy Section Weldment. *Weld. J.* **2017**, *96*, 203s–219s.
10. Abe, F.; Tabuchi, M.; Tsukamoto, S.; Shirane, T. Microstructure evolution in HAZ and suppression of type IV fracture in advanced ferritic power plant steels. *Int. J. Press. Ves. Pip.* **2010**, *87*, 598–604. [CrossRef]
11. Zhao, L. Experimental investigation of specimen size effect on creep crack growth behavior in P92 steel welded joint. *Mater. Des.* **2014**, *57*, 736–743. [CrossRef]
12. Jovanović, M.; Čamagić, I.; Sedmak, S.; Živković, P.; Sedmak, A. Crack Initiation and Propagation Resistance of HSLA Steel Welded Joint Constituents. *Struc. Integr. Life* **2020**, *20*, 11–14.
13. Čamagić, I.; Jović, S.; Radojković, M.; Sedmak, S.; Sedmak, A.; Burzić, Z.; Delamarian, C. Influence of Temperature and Exploitation Period on the Behaviour of a Welded Joint Subjected to Impact Loading. *Struc. Integr. Life* **2016**, *16*, 179–185.
14. Čamagić, I.; Sedmak, S.; Sedmak, A.; Burzić, Z.; Todić, A. Impact of Temperature and Exploitation Time on Plane Strain Fracture Toughness, K_{Ic}, in a Welded Joint, I. *Struc. Integr. Life* **2017**, *17*, 239–244.
15. Čamagić, I.; Sedmak, A.; Sedmak, S.; Burzić, Z. Relation between impact and fracture toughness of A-387 Gr. B welded joint, 25th International Conference on Fracture and Structural Integrity. *Procedia Struct. Integr.* **2019**, *18*, 903–907. [CrossRef]
16. Čamagić, I.; Sedmak, S.; Sedmak, A.; Burzić, Z.; Marsenić, M. Effect of temperature and exploitation time on tensile properties and plain strain fracture toughness, K_{Ic}, in a welded joint, IGF Workshop "Fracture and Structural Integrity". *Procedia Struct. Integr.* **2018**, *9*, 279–286. [CrossRef]
17. Čamagić, I.; Sedmak, S.; Sedmak, A.; Burzić, Z.; Aranđelović, M. The impact of the temperature and exploitation time on the tensile properties and plain strain fracture toughness, K_{Ic} in characteristic areas of welded joint. *Frat. Ed Integrita Strut.* **2018**, *46*, 371–382.
18. Čamagić, I.; Vasić, N.; Ćirković, B.; Burzić, Z.; Sedmak, A.; Radović, A. Influence of temperature and exploitation period on fatigue crack growth parameters in different regions of welded joints. *Frat. Integrita Strut.* **2016**, *36*, 1–7.
19. Jovanović, M.; Čamagić, I.; Sedmak, A.; Burzić, Z.; Sedmak, S.; Živković, P. Analysis of SA 387 Gr. 91 welded joints crack resistance under static and impact load. *Procedia Struct. Integr.* **2021**, *31*, 38–44. [CrossRef]
20. *SRPS EN ISO 9016:2013: Destructive Tests on Welds in Metallic Materials—Impact Tests—Test Specimen Location, Notch Orientation and Examination*; Serbian Institute for Standarisation: Belgrade, Serbia, 2013.
21. Grabulov, V.; Burzić, Z.; Momčilović, D. *Significance of Mechanical Testing for Structural Integrity*; IFMASS: Belgrade, Serbia, 2008.
22. Mijatović, T.; Manjgo, M.; Burzić, M.; Čolić, K.; Burzić, Z.; Vuherer, T. Structural integrity assessment from the aspect of fracture mechanics. *Struct. Integr. Life* **2019**, *19*, 121–124.
23. *ASTM E 1820–99a: Standard Test Method for Measurement of Fracture Toughness, Annual Book of ASTM Standards*; ASTM International: West Conshohocken, PA, USA, 1999.
24. *BS 7448-Part 2: Fracture Mechanics Toughness Tests—Methods for Determination of K_{Ic}, Critical CTOD and Critical J Values of Welds in Metallic Materials*; BBI: Grand Forks, ND, USA, 1997.

 *materials* 

Article

Cleavage Fracture of the Air Cooled Medium Carbon Microalloyed Forging Steels with Heterogeneous Microstructures

Gvozden Jovanović [1,*], Dragomir Glišić [2], Stefan Dikić [2,*], Nenad Radović [2] and Aleksandra Patarić [1]

[1] Institute for Technology of Nuclear and Other Mineral Raw Materials, Metallurgical and Environmental Engineering, Bulevar Franše d'Eperea 86, 11000 Belgrade, Serbia; a.pataric@itnms.ac.rs

[2] Faculty of Technology and Metallurgy, University of Belgrade, Karnegijeva 4, 11120 Belgrade, Serbia; gile@tmf.bg.ac.rs (D.G.); nenrad@tmf.bg.ac.rs (N.R.)

[*] Correspondence: g.jovanovic@itnms.ac.rs (G.J.); sdikic@tmf.bg.ac.rs (S.D.)

Abstract: Cleavage fracture of the V and Ti-V microalloyed forging steels was investigated by the four-point bending testing of the notched specimens of Griffith-Owen's type at $-196\,°C$, in conjunction with the finite element analysis and the fractographic examination by scanning electron microscopy. To assess the mixed microstructure consisting mostly of the acicular ferrite, alongside proeutectoid ferrite grains and pearlite, the samples were held at $1250\,°C$ for 30 min and subsequently cooled instill air. Cleavage fracture was initiated in the matrix under the high plastic strains near the notch root of the four-point bending specimens without the participation of the second phase particles in the process. Estimated values of the effective surface energy for the V and the Ti-V microalloyed steel of $37\ Jm^{-2}$ and $74\ Jm^{-2}$, respectively, and the related increase of local critical fracture stress were attributed to the increased content of the acicular ferrite. It was concluded that the observed increase of the local stress for cleavage crack propagation through the matrix was due to the increased number of the high angle boundaries, but also that the acicular ferrite affects the cleavage fracture mechanism by its characteristic stress–strain response with relatively low yield strength and considerable ductility at $-196\,°C$.

Keywords: cleavage fracture stress; medium carbon forging steel; microalloyed steel; acicular ferrite; heterogeneous microstructure

Citation: Jovanović, G.; Glišić, D.; Dikić, S.; Radović, N.; Patarić, A. Cleavage Fracture of the Air Cooled Medium Carbon Microalloyed Forging Steels with Heterogeneous Microstructures. *Materials* **2022**, *15*, 1760. https://doi.org/10.3390/ma15051760

Academic Editor: Dražan Kozak

Received: 4 November 2021
Accepted: 30 November 2021
Published: 25 February 2022

Publisher's Note: MDPI stays neutral with regard to jurisdictional claims in published maps and institutional affiliations.

1. Introduction

Medium carbon microalloyed forging steels have been introduced into forging practice with the intent to replace costly procedures of quenching and tempering. Despite the fact they are high in strength, in some cases forging steels lack the desired impact toughness up to 40%, as shown in [1]. Forged steel parts are usually produced by heating to a high temperature and forged to the desired shape. In the case of the drilling rods for the petrol industry, the rod's end is heated with an induction heater to about $1200\,°C$ and forged into a suitable shape for subsequent machine working, as shown in Figure 1. Consequently, the neighboring part of the rod is heated above Ac_1 temperature by conduction, therefore forming a heat affected zone (HAZ), fully analogous to the HAZ in a welding process.

By continuous cooling from the forging temperatures, a range of heterogeneous microstructures can be formed. Depending on the cooling rates and the chemical composition, microstructure of the medium carbon microalloyed steel could consist of ferrite, perlite, bainite, acicular ferrite, martensite, and retained austenite [2,3]. The same heterogeneous structures can be formed in HAZ both in the case of welding and forging, except for the differences in the content of each microconstituent and the size proportions of the HAZ.

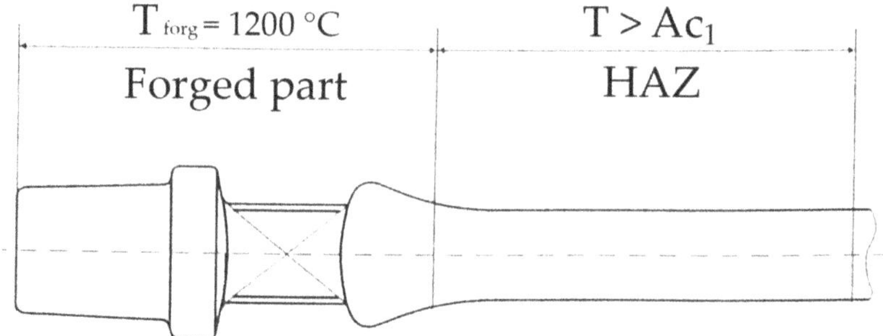

Figure 1. Schematic illustration of heat affected zone in the forged rod.

Previous research have noted the positive influence of acicular ferrite (AF) in the microstructure on the toughness of low carbon and welded pipeline steels [4–10]. AF is formed by bainitic reaction, but unlike bainite that grows as a sheaf of ferrite plates from the austenite grain boundary, AF nucleates intragranularly at inclusions, forming fine interlocking structure of laths and/or plates [11,12]. It is believed that AF increases toughness by forcing propagating crack to deflect at the high angle boundaries between the ferrite laths and plates [13]. Formation of AF is also observed in medium carbon microalloyed steels, coexisting with other microconstituents, primarily ferrite and pearlite, in continuously cooled samples from the temperature of austenitization or hot working [3,14]. Promising results in improving the toughness of the medium carbon microalloyed steels have been noted in those with a predominantly acicular ferrite structure [4,15–17]. For continuously cooled medium carbon microalloyed steels fracture studies were focused mainly on ferrite-pearlite and bainite structures [18,19], and it was found that in general, cleavage fracture was initiated or controlled by the broken coarse TiN particles in the zone of high stress in front of the notch in the four-point bending specimens [20–22]. Effective surface energy for those steels has been determined to be less than 50 Jm^{-2} at -196 °C [20,23]. However, little attention has been given to the cleavage fracture in continuously cooled medium carbon microalloyed steels with the predominantly acicular ferrite structure.

The aim of this work is to investigate the influence of a classical heterogeneous structure predominantly made up of acicular ferrite on the micromechanical mechanism of cleavage fracture for two different medium carbon microalloyed steels. An attempt is made to better understand how the control of a heterogeneous structure (i.e., acicular ferrite) influences fracture behavior, similar to approaches in the studies of other multiphase steels, such as TrIP (transformation induced plasticity), TwIP (twinning induced plasticity), or DP (dual phase) steels [24].

2. Materials and Methods

The chemical compositions of two commercial medium carbon microalloyed steels are given in Table 1. The Ti-V and V microalloyed steel were received as hot-rolled rods 22 mm and 19 mm in diameter, respectively. In order to eliminate rolling texture, the as-received rods were homogenized at 1250 °C for 4 h in an argon atmosphere, followed by quenching in oil at room temperature. In order to achieve predominantly acicular ferrite structure by continuous cooling, specimens were afterward reaustenitized at 1250 °C for 30 min in argon and then cooled at still air.

Table 1. Chemical composition of the steels (wt.%).

Steel	C	Si	Mn	P	S	Cr	Ni	Mo	V	Ti	Al	Nb	N
Ti-V	0.309	0.485	1.531	0.0077	0.0101	0.265	0.200	0.041	0.123	0.011	0.017	0.003	0.0228
V	0.256	0.416	1.451	0.0113	0.0112	0.201	0.149	0.023	0.099	0.002	0.038	0.002	0.0229

Metallographic specimens were cut from the rods in a transverse direction and prepared for light microscopy examination by grinding, polishing, and etching in 2% solution of nitric acid in ethanol. The microstructure of the steels was examined by light microscopy, and the micrographs were captured by using a digital camera. Quantitative analysis of the microstructures was performed using FIJI software [25]. A separate set of specimens was prepared for measurement of the previous austenite grain size (PAGS). In order to reveal austenite grain boundaries, specimens were reheated to 1250 °C for 15 min, water quenched, subsequently tempered at 450 °C for 24 h, cooled in still air to room temperature subsequently etched in a saturated solution of picric acid with 1 cm^3 of HNO$_3$. Average values of PAGS measured by the linear intercept of the grain boundary segregations network were 80 ± 10 μm for Ti-V steel and 100 ± 10 for V steel.

In order to investigate the cleavage fracture of the steels four-point bending (4PB) tests were carried out at the temperature of liquid nitrogen (-196 °C) with a constant crosshead speed of 0.1 mm min^{-1}. Four-point bending notched specimens are the same type as those used by Griffiths-Owen [26].

The finite element modeling (FEM) was performed using SIMULIA Abaqus software in order to calculate stresses and strains in the four-point bending (4PB) specimens. One-quarter of the specimen was modeled in three dimensions. Hexagonal eight-node element type with reduced integration was used (C3D8R) with geometric nonlinearity taken into account. The mesh was generated so that the size of the elements is gradually refined toward the notch tip, from 0.7 to 0.07 mm. Displacement controlled loading was applied with the maximum displacement that corresponds to the fracture load reached in the experiment. Elastic-plastic response of the steels was modeled by the elastic modulus and Poisson's ratio of 200 GPa and 0.28 [26], respectively, and by using the true stress–strain curves constructed by polynomial regression of the experimental data obtained by the uniaxial tension testing at -196 °C.

The stress–strain dependence was determined by uniaxial tensile testing in a liquid nitrogen bath at a constant crosshead speed of 0.1 mm min^{-1}, which provides an initial strain rate of the same order of magnitude as in the four-point bending test ($\approx 10^{-5}$ s^{-1}). Uniaxial tensile testing was performed using proportional cylindrical specimens 6 mm in diameter and 30 mm in gauge length (EN ISO 6892-1).

Fracture analysis was performed using a scanning electron microscope (SEM) equipped with an energy dispersive X-ray spectrometer (EDS). Fracture origins were determined by tracking the markings on the cleavage facets. Distance of the fracture initiation site from the notch root, X_0, and the size of the initial cleavage facet was measured at the SEM micrographs. The cleavage facets were approximated by an ellipse with major axes corresponding to the maximum and minimum ferret diameters of the facet (D_{max} and D_{min}, respectively).

The local cleavage fracture stress, σ_F^*, was determined from the FEM calculated maximum principal stress distribution at the distance of the cleavage initiation site from the notch root, X_0. Critical cleavage fracture stress, σ_F^*, and the first cleavage facet dimensions are related by Griffith's equation [18,19]:

$$\sigma_F^* = \sqrt{\frac{\pi \cdot E \cdot \gamma}{(1 - \nu^2) \cdot D_{eff}}} \tag{1}$$

where γ is the effective surface energy, D_{eff} is the effective diameter of the first cleavage facet, E is the modulus of elasticity, and ν is the Poisson's coefficient. In this paper, effective diameter, D_{eff}, was calculated by the following formulas [17,18]:

$$D_{eff} = \frac{D_{min}}{\phi^2} \frac{\pi^2}{4} \tag{2}$$

$$\phi = \frac{3\pi}{8} + \frac{\pi}{8}\left(\frac{D_{min}}{D_{eff}}\right)^2 \tag{3}$$

Based on Equation (1), the effective surface energy, γ, was determined from the plot of the local critical cleavage fracture stress versus the reciprocal square root of the first cleavage facet effective diameter, $\sigma_F^{*}\text{-}D_{eff}^{-1/2}$

3. Results and Discussion

3.1. Microstructure

Microstructures of the steels investigated, shown in Figure 2, consist of ferrite, pearlite, and acicular ferrite. The steel with V as the main microalloying addition ("V steel") is characterized by the continuous network of proeutectoid grain boundary ferrite (GBF) along the previous austenite grain boundaries bordered by the pearlite nodules (P). Grain boundary ferrite consists of both elongated allotriomorphs (GBA) and polygonal idiomorphs (GBI). Most of the previous austenite grain interior is occupied by acicular ferrite (AF), mainly separated from the GBF by the perlite (P), as can be observed in Figure 2a,c. Unlike the V steel, in the structure of the steel microalloyed with Ti and V ("Ti-V steel") proeutectoid ferrite grains and pearlite are sparse and discontinuous, while acicular ferrite occupies an almost complete volume of the previous austenite grain (Figure 2b,d). Estimated volume fractions for V steel are 70% of AF and 20% of P, while for steel, it is 96% AF and only 2% P, and the rest is GBF.

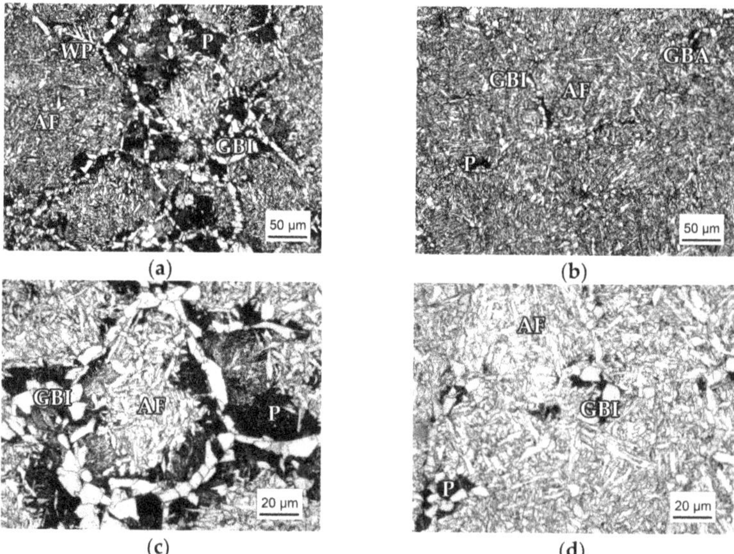

(a)   (b)   (c)   (d)

Figure 2. Optical micrographs of the medium carbon V and Ti-V microalloyed steels samples air-cooled from the austenitization temperature of 1250 °C: (**a**) Overall microstructure of the V microalloyed steel; (**b**) overall microstructure of the Ti-V microalloyed steel; (**c**) details of the V steel microstructure;(**d**) details of the Ti-V steel microstructure. P-pearlite, GBI-grain boundary idiomorphs, GBA-grain boundary allotriomorphs, AF-acicular ferrite, WP—Widmanstätten side plates.

Small isolated islands of pearlite alongside a discontinuous network of GBF indicate markedly higher hardenability of the Ti-V steel. Hardenability in terms of retardation of diffusional transformation of austenite can be rationalized in terms of the chemical compositions of the steels. Besides higher carbon content, the higher hardenability of Ti-V steel is attributed to the higher content of substitutional alloying elements—Cr, Mo, and Ni. Manganese content, which is known to have a strong retarding effect on the diffusional decomposition of austenite [27], is essentially equal for both steels. However, a possible influence of free excess vanadium atoms in austenite solid solution that are not tied in V(C,N) particles should be taken into consideration. Vanadium atoms in solid solution increase hardenability by segregating to austenite grain boundaries, rendering them energetically less suitable for nucleation of proeutectoid ferrite [6,28]. The distribution of the microalloying elements together with the temperatures for the complete dissolution of VN and VC particles is given in Table 2. Considering temperatures for complete dissolution calculated using equations for solubility products [29]:

$$log[V][N] = -7840/T_{VN} + 3.02 \tag{4}$$

$$log[V][C] = -9500/T_{VC} + 6.72 \tag{5}$$

it follows that VN and VC carbides were completely dissolved at the austenitization temperature of 1250 °C. We can assume that the total amount of titanium is precipitated as TiN particles due to low solubility even at the austenitization temperature [19].

Table 2. Redistribution of the Ti, V, and N elements between particles and solid solution, and temperatures for the complete dissolution of VN and VC precipitates.

Steel	[Ti] [ppm]	[N] [ppm]	[N]TiN [ppm]	[N]VN [ppm]	[V] [wt.%]	[V]VN [wt.%]	[V]excess [wt.%]	T_{VN} [°C]	T_{VC} [°C]
Ti-V	110	228	32	196	0.123	0.071	0.052	1117	894
V	20	229	6	223	0.099	0.081	0.018	1108	869

The distribution of the microalloying elements was calculated by taking the stoichiometric ratios in TiN and VN particles of Ti:N = 3.4, and V:N = 3.6, with the assumption that upon cooling from the austenitization temperature, precipitation of VN particles takes precedence over the precipitation of VC particles. From these considerations, it follows that higher concentrations of free vanadium atoms can be expected in Ti-V steel, and consequently, that it would impose a stronger suppressing effect on diffusional transformations of the austenite.

The influence of the prior austenite grain size (PAGS) on hardenability is related to the number density of potential grain boundary nucleation sites [30]. The addition of Ti in the Ti-V steel contributes to grain size control and austenite grain refinement [13,31]. Austenite grain size in both steels examined was relatively large, about 80 μm in Ti-V steel and 100 μm in V steel, which is favorable for the intragranular nucleation. The addition of Ti was in order to prevent grain growth of austenite at forging temperature, i.e., no influence on hardness was expected. The difference in PAGS did not impose a noticeable effect on the hardenability of the steels; thus, higher hardenability of the Ti-V steel implies the dominant influence of the alloying elements in solid solution, especially free excess V.

Transformation on continuous cooling from the austenitization temperature begins with proeutectoid ferrite formation at austenite grain boundaries, advances by the formation of the pearlite, and finishes by bainitic reaction at lower temperatures when diffusional transformations are no longer possible. Nucleation of the bainite at the austenite grain boundaries is precluded by the presence of grain boundary ferrite grains, mostly in V steel, or due to segregation of alloying elements, presumably free excess V in Ti-V steel, and therefore intragranular nucleation of acicular ferrite takes place. Intragranular nucleation is promoted by the presence of second phase particles suitable as ferrite nucleation sites. VN particles precipitated at MnS inclusions seem particularly effective as intragranular ferrite nucleation site [32,33]. Considering the chemical composition of the steels investigated, complex VN/MnS particles could be expected as the main acicular ferrite nucleation site, although other particles with MnS core, such as CuS and TiN should not be excluded [20,34–37].

AF is considered as an effective barrier for crack propagation due to the high density of high angle boundaries by forcing propagating cracks to change direction frequently, thus increasing the overall energy needed for fracture [6]. However, previous investigations established two distinct AF morphologies, generally defined in analogy to bainite, depending on the temperature of isothermal transformation, as upper and lower AF [38]. Lower AF formed at temperatures around 400 °C is depicted by the sheaves of nearly parallel plates or laths with similar crystallographic orientation, while upper AF formed at around 450 °C is characterized by the fine interlocking structure of ferrite plates/laths with high crystallographic misorientation [4,12,39]. Therefore, it seems rational to consider acicular ferrite obtained by continuous cooling as a mixture of both isothermal types, although it cannot be easily discerned at the light microscopy level.

3.2. Fractography

Macroscopic V-shaped markings at the fracture surface ("chevron markings") points to the origin of the fracture located near the notch root (Figure 3a,b). The fracture surface is characterized by the fine irregular cleavage facets (Figure 3c,d) alongside with the islands of coarse facets. These isolated coarse facets are also irregularly shaped and more often present in the Vsteel than in the Ti-Vsteel (Figure 3c,d). Observed fracture surface features could be related to the underlying structure consisting of various ratios of grain boundary ferrite, pearlite, and acicular ferrite. In that manner, large facets may be correlated with the fractured grain boundary ferrite grains and/or coarse pearlite nodules, while small irregular facets correspond to the fractured AF plates.

The appearance of the fracture surface reflects the underlying microstructures of the two steels examined. Therefore, in the Ti-V steel with the predominantly AF structure and lower volume fraction of GBF and pearlite, there are fewer coarse cleavage facets, which are generally smaller in size than in the V steel specimens. An example of coarse facets or a group of facets is shown in Figure 3c,d for the V and Ti-V steel, respectively. At small facets, fracture marks are barely visible and could be considered as feather markings rather than river lines, which makes them difficult to trace back to the origin of the fracture. Pronounced tearing lines bordering fine facets seem to indicate an effect of the high-angle crystallographic misorientation between individual acicular ferrite plates, forcing propagating crack frequently to deflect. An example of the group of coarse facets separated by the ridge that represents the boundary between adjacent grains with tilted crystal orientations is shown in Figure 3f. Such coarse facets could be associated not only with individual GBF and pearlite but also with the aggregates of ferrite and pearlite with similar or the same crystallographic orientation [40,41]. However, cleavage fracture traces bypass the facets in this example, shown in Figure 3f, and therefore could not be the origin of the cleavage fracture. Additionally, having in mind that continuously cooled structures of the medium carbon microalloyed steels may contain both upper and lower acicular ferrite, it could be assumed that some of the coarser facets correspond to the sheaves of acicular ferrite plates with similar crystallographic orientation. In this respect, previous studies have indicated that microstructural units controlling cleavage fracture in an acicular ferrite structure could be a group of ferrite plates with crystallographic misorientation below 15° [13,42].

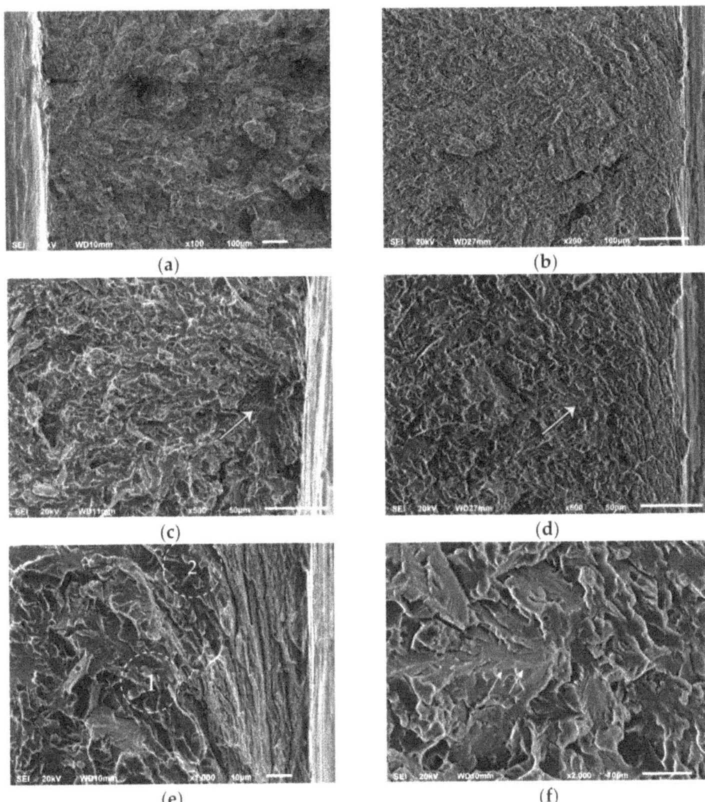

Figure 3. SEM micrographs of the fracture surfaces: (**a**) Macroscopic markings pointing to the fracture initiation site in V steel; (**b**) macroscopic markings pointing to the fracture initiation site in Ti-V steel; (**c**) coarse facets near the fracture origin in V steel sample; (**d**) fracture origin in Ti-V steel sample and an example of the tilted boundary between two coarse cleavage facets nearby; (**e**) typical cleavage initiation site in V steel, marked with number 1, (**f**) coarse facets separated by the boundary between the two grains with tilted crystallographic orientation near the fracture origin in Ti-V steel.

In both steels fracture originates from the area almost at the notch root. There are not any fractured second phase particles or inclusions at the cleavage fracture origins in any of the samples of both steels. SEM micrograph with EDS in Figure 4 shows spherical MnS-based complex particle found near the cleavage initiation site. This particle is not broken, fine cleavage markings on the surrounding facets bypass it, and therefore not related to the fracture initiation. By following river lines and feather markings, a facet or a group of facets at the initiation sites near the notch root were discovered in both steels. Distances measured from the notch root to the cleavage initiation sites are between 14 and 56 μm for V steel specimens and 56 to 131 μm for the Ti-V steel specimens (Table 3). According to the distribution of the maximum principal stress, σ_{11}, calculated by the FEM, shown in Figure 5, peaks of the maximum principal stress reach approximately 2300 MPa and 2500 MPa for V steel and Ti-V steel, respectively. Cleavage fracture in microalloyed steels with ferrite-pearlite, bainite, or martensite structures is generally initiated by the fracture of a coarse second phase particle. In the vast majority of the cases, it was TiN particle, 26 μm in diameter, in the zone of the peak value of the maximum principal stress [18,20,23,43,44]. It could be concluded that in this case, with the predominantly AF structure, an alternative mechanism of cleavage initiation was active. A similar case of cleavage initiation was also found near the notch of the 4PB. Therefore, it could be concluded that the local stress did not attain the critical level for the fracture of coarse TiN particles. Furthermore, from the graphs shown in Figure 5, it is clear that all the fracture origins are located within the narrow zone of high plastic deformations.

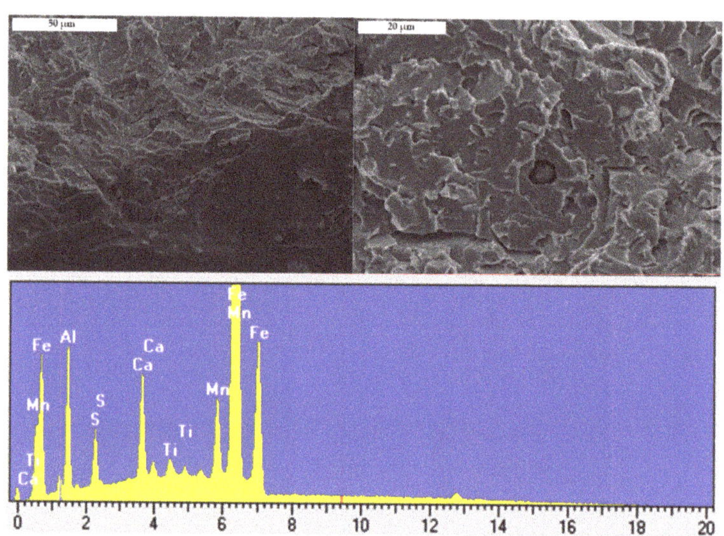

Figure 4. SEM micrographs and the EDS spectra of the particle found near the notch of the 4PB specimen.

Table 3. Critical parameters for the cleavage fracture initiation for V and Ti-V steel samples.

V Steel Sample	σ_F [MPa]	σ_F/σ_0	σ_{1max} [MPa]	X_0 [μm]	σ_F^* [MPa]	ε_{pc}	$D_{max} \times D_{min}$ [μm]	D_{eff} [μm]	γ [Jm^{-2}]
1	1193	1.54	2645	32	1419	0.493	36.8 × 15.5	24.5	72.3
				53	1471	0.450	12.8 × 17.3	11.9	37.7
				19	1400	0.634	23.7 × 9.8	15.6	44.7
2	1268	1.64	2671	19	1400	0.634	21.0 × 11.3	16.7	48.0
				28	1414	0.616	29.3 × 24.1	28.5	83.7
3	1049	1.35	2595	15	1398	0.369	41.0 × 8.1	14.0	40.1
				14	1396	0.370	8.6 × 6.8	8.3	23.6
4	1152	1.49	2640	56	1479	0.409	34.7 × 8.2	14.0	45.1
Ti-V Steel Sample	σ_F [MPa]	σ_F/σ_0	σ_{1max} [MPa]	X_0 [μm]	σ_F^* [MPa]	ε_{pc}	$D_{max} \times D_{min}$ [μm]	D_{eff} [μm]	γ [Jm^{-2}]
1	759	0.85	2139	78	1739	0.060	20.4 × 18.3	20.2	89.7
2	934	1.04	2402	59	1751	0.152	18.8 × 12.7	17.0	76.5
				56	1740	0.154	22.3 × 13.6	19.1	85.0
3	904	1.01	2343	112	1934	0.092	17.2 × 15.3	17.0	93.4
				131	1983	0.082	22.1 × 17.6	21.3	123.0
4	950	1.06	2430	71	1792	0.156	35.4 × 12.3	20.2	95.2
				65	1770	0.161	29.1 × 11.8	18.8	86.6

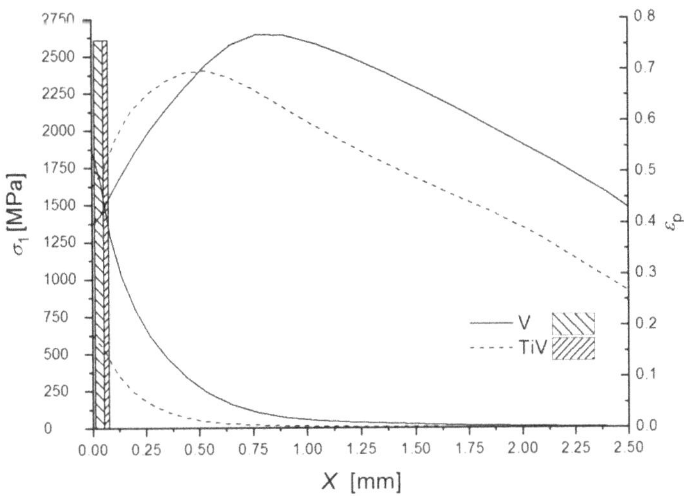

Figure 5. Maximum principal stress and the plastic strain distribution along the distance from the notch root X_0 of the 4PB samples for the V and Ti-V microalloyed steel at the cleavage fracture initiation. Shaded boxes indicate the zone of the cleavage fracture initiation.

Cleavage fracture initiation by the Smith's mechanism, involving fracture of the cementite plate at the ferrite grain boundary [45,46] also does not seem probable in this case, at the stress levels present in the vicinity of the notch. Taking the value of effective surface energy for cementite of 9 Jm^{-2} [47] and by using Griffith's equation for the thru-thickness crack:

$$\sigma_F^* = \sqrt{\frac{4E\gamma}{\pi(1-\nu)C}} \tag{6}$$

it can be calculated that the highest stress values at the locations of the fracture initiation given in Table 3, were not sufficient for the propagation of the cracks nucleated at the cementite plates thinner than approximately 0.7 μm. This consideration leads to an assumption that initial microcracks were formed by the rupture of the grains under the high plastic deformations. In fractographs in Figure 3d,e, on the right side, many ruptures formed at the notch root are seen. Facet in Figure 3e marked as number 2 was formed almost at the notch root but had not been a part of the main propagating crack front. Bearing in mind the role of the high plastic strains, a mechanism of damage accumulation in pearlite by fracture of multiple cementite lamellas (Miller–Smith mechanism) could be considered as a plausible mechanism of the cleavage crack nucleation, in analogy to the cleavage fracture initiation in the pearlitic steels [48,49]. It could be assumed that microcracks formed at the pearlite nodules would easily propagate through the low angle boundary with neighboring proeutectoid ferrite. A relatively large microcrack formed by this mechanism could easily propagate at relatively low stresses near the notch tip.

Plastic yielding is confined within the narrow zone, about 1.5 mm from the notch root for V steel, and about 0.75 mm for Ti-V steel (Figures 5 and 6). Plastic strains in V steel reach values as high as 0.6, while in Ti-V steel they are limited to much lower values, up to approximately 0.15 (Figure 5, Table 3). In the same manner, the values of the plastic strains at the cleavage fracture initiation sites, ε_{pc}, given in Table 3, are also considerably different for the two steels examined. Observed behavior could be related to the stress–strain response of the steels, presented by the fitted experimental true stress–strain curves at $-196\,^\circ$C, shown in Figure 7 [50], which had been used as a model of the mechanical properties in FEM. Ti-V steel exhibits considerable ductility at $-196\,^\circ$C in comparison with V steel sample. On the other hand, V steel is characterized by the higher strain-hardening rate, leading to the observed distribution of the plastic strain around the notch and formation of the narrow plastic zone with steeper strain increase toward the notch in V steel than in Ti-V steel four-point bending sample. The deformation behavior of the steels tested could be related to its microstructure and the content of the individual microconstituents. It is known that the increase of the pearlite volume fraction leads to the increase of the strength and strain-hardening rate [51,52].

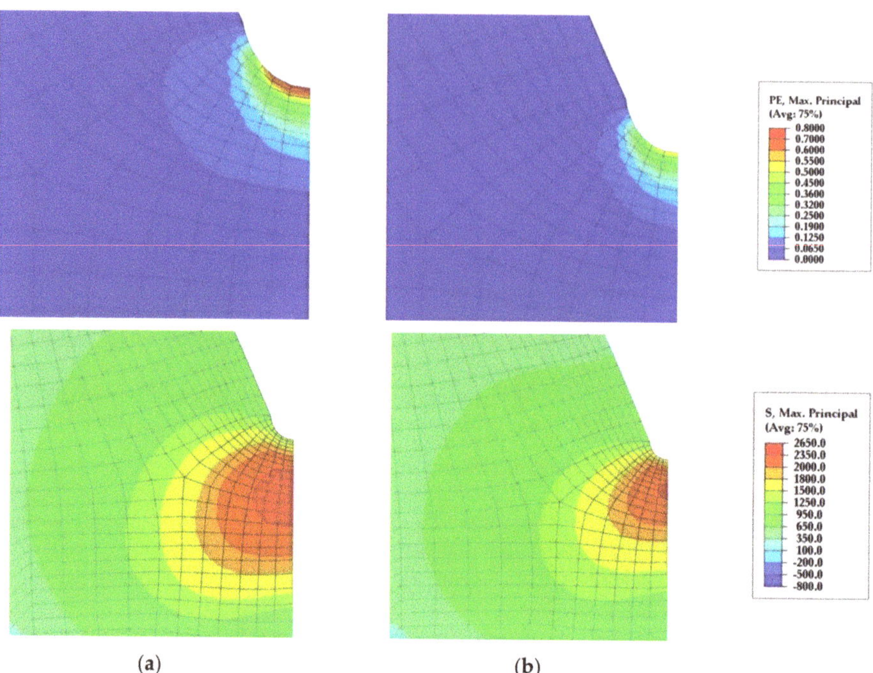

Figure 6. Contours of the plastic strain (above) and the maximum principal stress(below) distribution in front of the notch root of the 4PB specimen at the cleavage fracture initiation from the same viewpoint in FEMfor the (**a**) V steel; (**b**) Ti-V steel.

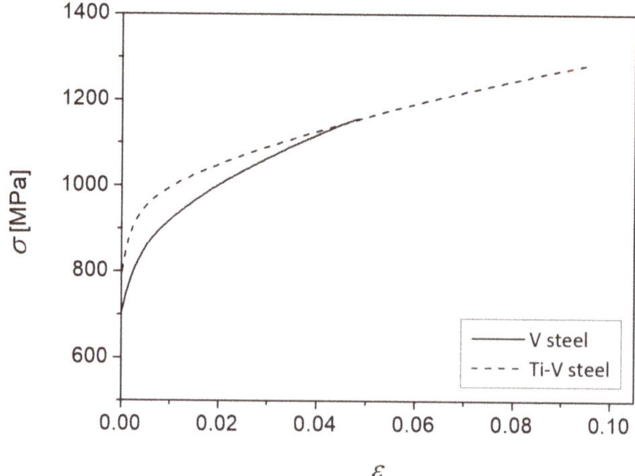

Figure 7. Fitted experimental tensile true stress–true strain curves of the V steel and the Ti-V steel samples tested at −196 °C.

Further, both steels exhibit a gradual- or continuous-yielding effect, which manifests as the relatively low YS, quantified as the $R_{p0.2}$ value, followed by the steep stress increase at low strains (Figure 7), and the consequently lower YS/UTS ratio. This behavior is characteristic of the steels with bainitic or AF structure that contains high density of mobile dislocations due to a displacive nature of transformation [53,54]. Furthermore, another possible cause for the observed mechanical response could be the presence of the retained austenite, characteristic of the medium-carbon microalloyed steels with bainitic or AF structure, related to the incomplete reaction phenomenon [55]. It could be in particular related to the observed tensile elongation of the Ti-V steel specimens at such low temperatures (Figure 7). In conclusion, the observed distribution of the strain in the four-point bending specimen, calculated by FEM, was a consequence of the mechanical behavior of the continuously cooled steel microstructures with the predominant content of AF and the contribution of the various amount of pearlite and GBF depending on the cooling rate as well as the chemical composition of the microalloyed steels.

Having in mind that the origins of the cleavage fracture were found exclusively in the narrow area at the notch root, it could be assumed that high plastic deformation led to the formation of the microcracks by fracture of the microconstituents with low ductility, primarily the pearlite, which resides in the microstructure between the AF and GBF. A rupture of GBF grains by the Smith's mechanism due to low ductility at $-196\,^\circ$C should also be taken into account. The size of the facets at the cleavage origin, determined as the effective diameter, D_{eff}, shown in Table 3, is rather similar for V and Ti-V steel, with the average values of 17 μm and 19 μm, respectively. A similar size of the first facet's diameter regardless of the size and the volume fraction of pearlite and GBF in the two steels may imply a possibility that initial microcracks form also in AF, presumably at sheaves of ferrite plates with similar crystallographic orientation (lower AF).

However, distances of the cleavage origins from the notch in Ti-V steel 4PB specimens are somewhat larger, and therefore the local cleavage stress is higher. Although nominal fracture stress, σ_F, is lower for the Ti-V steel (Table 3, columns 2 and 3), the local stress at the initiation site is higher than in the V steel. Therefore, when calculated using Griffith's equation for the circular crack (Equation (1)), the value of the effective fracture surface energy is noticeably higher for the Ti-V steel (Table 3, column 10).

Considering linear dependence of the local critical fracture stress, σ_F^{*}, on the reciprocal square root of the initial microcrack effective diameter, $D_{eff}^{-1/2}$, in accordance with the Griffith's equation (Equation (1)) lines have been drawn in the graph in Figure 8 in order to calculate the values of the effective surface energy for cleavage fracture, γ. Experimental points lie in the range between 37 Jm^{-2} and 82 Jm^{-2} for the V steel and between 74 Jm^{-2} and 122 Jm^{-2} for the Ti-V steel. From the line drawn just below the lowest experimental point, an assumed upper value of the true effective surface energy for cleavage fracture, γ, is calculated. Therefore, the values of 37 Jm^{-2} and 74 Jm^{-2} could be adopted for the V and Ti-V steel, respectively. One of the experimental points for V steel deviates to a large degree, giving the unrealistic value of 24 Jm^{-2}, and therefore it was excluded from the analysis. This point matches the unusually small facet diameter in sample 3 in Table 3. Having in mind that cleavage initiation, in this case, is not related to the fracture of particles but to the plastic deformation at the notch root, the calculated values of the effective surface energy for cleavage fracture could be considered as the values for the propagation of the crack through the grain boundaries, $\gamma = \gamma_{mm}$. In that manner, it indicates the role of the fine interlocking structure of AF in the cleavage fracture mechanism. The value of the effective fracture surface energy of 37 Jm^{-2} for the V steel with a mixed microstructure consisting of about 70% of AF, and the rest being mostly the pearlite and GBF, is comparable to the values for the medium-carbon microalloyed steels with the ferritic-pearlitic and bainitic structures found in the literature [20,23].

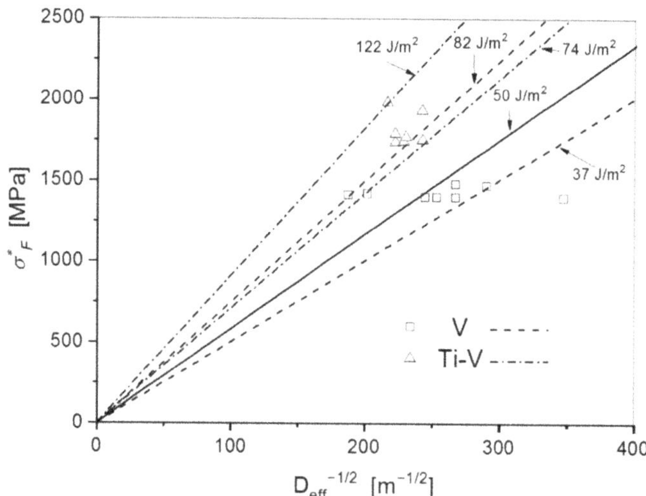

Figure 8. Values of the local critical fracture stress plotted against the reciprocal square root of the first cleavage fracture facet's effective diameter and the corresponding values of the effective fracture surface energy.

However, the value 74 Jm^{-2} for the Ti-V steel is considerably higher, and it could be assumed that the observed increase in the effective surface energy was due to the higher density of high-angle boundaries of the AF structure. Additionally, another rather indirect effect of the AF through its influence on the stress–strain behavior and resultant stress and strain distribution in front of the notch in 4PB specimen could be considered.

First, due to the characteristic high dislocation density, the observed low yield strength or the "gradual yielding" effect, alongside with the relatively high ductility at low temperatures (Figure 7), induce high plastic strains at relatively modest stress levels near the notch, which cause the fracture of the favorably oriented pearlite nodules or coarse eutectoid ferrite grains.

Formed microcracks are large enough to trigger the cleavage fracture at low stresses near the notch root. At the same time, peak stress at the distance X_0, 400–500 μm from the notch root, had not been sufficient for the cleavage initiation by fracture of the coarse TiN particles. Taking the value of γ_{pm} =7 Jm^{-2}for the fracture of the coarse TiN particles [40,56], calculation using Griffith's equation (Equation (1)) gives the value of approximately 2500 MPa for the grain sizes smaller than 5.45 μm for V steel and 7.85 μm for Ti-V steel. Peak values of the maximum principal stress, σ_{1max}, for the V steel samples are in the range 2595–2671 MPa, while for the Ti-V steel samples are considerably lower, from 2139–2430 MPa (Table 3). Therefore, it follows that the peak values of the maximum principal stress, σ_{1max}, in this work was lower than the stress needed for the fracture of the coarse TiN particles in Ti-V steel. As regards V steel it could not be expected a significant number of large TiN particles considering the low content of Ti (Table 1). However, in previous researches, it was noticed that TiN particles could not trigger the cleavage fracture when embedded in a ductile and fine grained matrix [47,57]. The results in this investigation confirm that TiN remains neutral regarding the cleavage fracture initiation in the fine interlocking structure of acicular ferrite. In conclusion, the large cracks formed in the zone of high plastic deformation trigger the cleavage fracture before the critical conditions are met for the cleavage initiation by fracture of the coarse TiN particles in the zone of peak stress.

Furthermore, it also indicates the role of the pearlite and ferrite in the cleavage fracture mechanism of the microalloyed steels with predominantly AF structure as potential sites for the formation of microcracks. Therefore, a lower number of perlite nodules and polygonal ferrite grains in Ti-V steel implicate fewer sites for cleavage initiation. It could also explain somewhat larger distances of the cleavage fracture origin from the notch in Ti-V steel. Conversely, a larger number of pearlite nodules and proeutectoid ferrite grains in V steel represent a larger number of potential sites for the cleavage fracture initiation at lower stress levels near the notch root. While the fine interlocking structure of AF contributes to the increase of the local critical fracture stress, low-ductility coarse microconstituents–pearlite, GBF, or ferrite-pearlite aggregates with the same crystallographic orientation decrease the critical cleavage fracture stress, considering that they are suitable sites for cleavage fracture initiation, at relatively low stresses, in the medium carbon microalloyed steels with predominantly AF structure.

4. Conclusions

1. Cleavage fracture of air-cooled medium carbon microalloyed steels with predominantly acicular ferrite microstructure is initiated by the fracture of the coarse microstructural units, ferrite, and pearlite, under the plastic deformation at the notch root of the four-point bending specimen. In this case, fracture of the second phase particles is not involved in the cleavage fracture initiation process.
2. Peak stress in front of the notch root of the four-point bending specimen is insufficient for the cleavage fracture initiation by fracture of coarse TiN particles in the microstructure consisting predominantly of acicular ferrite due to its fine-grained structure with a high density of high-angle boundaries and relatively low strength and considerable ductility at liquid nitrogen temperature, comparing to the ferritic-pearlitic structure.
3. While the coarse ferrite grains and pearlite nodules govern the cleavage fracture initiation, stress and strain distribution in the four-point bending specimen are dominated by the acicular ferrite and its deformational characteristics. It is assumed that mechanical properties were also contributed by the retained austenite present in the acicular ferrite structure of the steel with higher carbon content and the Ti-V microalloying addition.
4. Estimated values of the effective surface energy for the V steel with about 70% of the acicular ferrite of 37 Jm^{-2}, and for the Ti-V steel with the acicular ferrite content as high as 96% of 74 Jm^{-2} are considered as the effective energy for the propagation of the crack through the grain boundaries, and therefore directly relate to the effect of the fine interlocking structure of acicular ferrite in the cleavage fracture mechanism.

Author Contributions. Conceptualization, D.G. and N.R.; methodology, D.G. and N.R.; validation, N.R. and A.P.; formal analysis, G.J., D.G. and A.P.; investigation, G.J. and S.D.; resources, N.R. and A.P.; data curation, D.G. and G.J.; writing—original draft preparation, G.J.; writing—review and editing, D.G. and S.D.; visualization, G.J. and S.D.; supervision, N.R.; project administration, N.R.; funding acquisition, D.G. All authors have read and agreed to the published version of the manuscript.

Funding: This work was supported by the Ministry of Education, Science and Technological Development of the Republic of Serbia (Contract No. 451-03-9/2021-14/200135). This work was carried out as partial fulfillment of the requirements for the PhD degree of Jovanović Gvozden at the University of Belgrade—Faculty of Technology and Metallurgy.

Data Availability Statement: The data presented in this study are available on request from the corresponding author. The data are not publicly available.

References

1. Engineer, S.; Huchtemann, B. Review and Development of Microalloyed Steels for Forgings, Bars and Wires. In Proceedings of the Second International Symposium on Microalloyed Bar & Forging Steel, Golden, CO, USA, 8–10 July 1996; pp. 61–78.
2. Drobnjak, D.; Koprivica, A. Microalloyed bar and forging steels. In Proceedings of the Second International Symposium on Microalloyed Bar & Forging Steel, Golden, CO, USA, 8–10 July 1996; pp. 93–107.
3. Glišić, D.; Radović, N.; Koprivica, A.; Fadel, A.; Drobnjak, D. Influence of reheating temperature and Vanadium content on transformation behavior and mechanical properties of medium carbon forging steels. *ISIJ Int.* **2010**, *50*, 601–606. [CrossRef]
4. Ishikavaa, F.; Takahashi, T. The Formation ot Intragranular Steels for Hot-torging Plates in Medium-carbon and Its Effect on the Toughness. *ISIJ Int.* **1995**, *35*, 1128–1133. [CrossRef]
5. Babu, S.S. The mechanism of acicular ferrite in weld deposits. *Curr. Opin. Solid State Mater. Sci.* **2004**, *8*, 267–278. [CrossRef]
6. He, K.; Edmonds, D.V. Formation of acicular ferrite and influence of vanadium alloying. *Mater. Sci. Technol.* **2002**, *18*, 289–296. [CrossRef]
7. Dong, J.; Liu, C.; Liu, Y.; Zhou, X.; Guo, Q.; Li, H. Isochronal phase transformation of Nb–V–Ti microalloyed ultra-high strength steel upon cooling. *Fusion Eng. Des.* **2017**, *125*, 423–430. [CrossRef]
8. Xiong, Z.; Liu, S.; Wang, X.; Shang, C.; Li, X.; Misra, R.D.K. The contribution of intragranular acicular ferrite microstructural constituent on impact toughness and impeding crack initiation and propagation in the heat-affected zone (HAZ) of low-carbon steels. *Mater. Sci. Eng. A* **2015**, *636*, 117–123. [CrossRef]
9. Peng, Y.; Chen, W.; Xu, Z. Study of high toughness ferrite wire for submerged arc welding of pipeline steel. *Mater. Charact.* **2001**, *47*, 67–73. [CrossRef]
10. Linaza, M.A.; Romero, J.L.; Rodriguez-Ibabe, J.M.; Urcola, J.J. Improvement of fracture toughness of forging steels microalloyed with titanium by accelerated cooling after hot working. *Scr. Metall. Et Mater.* **1993**, *29*, 1217–1222. [CrossRef]
11. Garcia-Mateo, C.; Capdevila, C.; Caballero, F.G.; de Andrés, C.G. Influence of V precipitates on acicular ferrite transformation part 1: The role of nitrogen. *ISIJ Int.* **2008**, *48*, 1270–1275. [CrossRef]
12. Díaz-Fuentes, M.; Gutiérrez, I. Analysis of different acicular ferrite microstructures generated in a medium-carbon molybdenum steel. *Mater. Sci. Eng. A* **2003**, *363*, 316–324. [CrossRef]
13. Díaz-Fuentes, M.; Iza-Mendia, A.; Gutiérrez, I. Analysis of different acicular ferrite microstructures in low-carbon steels by electron backscattered diffraction. Study of their toughness behavior. *Metall. Mater. Trans. A* **2003**, *34*, 2505–2516. [CrossRef]
14. Rasouli, D.; Asl, S.K.; Akbarzadeh, A.; Daneshi, G.H. Effect of cooling rate on the microstructure and mechanical properties of microalloyed forging steel. *J. Mater. Process. Technol.* **2008**, *206*, 92–98. [CrossRef]
15. Cao, Z.Q.; Bao, Y.P.; Xia, Z.H.; Luo, D.; Guo, A.M.; Wu, K.M. Toughening mechanisms of a high-strength acicular ferrite steel heavy plate. *Int. J. Miner. Metall. Mater.* **2010**, *17*, 567–572. [CrossRef]
16. Balart, M.J.; Davis, C.L.; Strangwood, M. Observations of cleavage initiation at (Ti,V)(C,N) particles of heterogeneous composition in microalloyed steels. *Scr. Mater.* **2004**, *50*, 371–375. [CrossRef]
17. Alexander, D.J.; Bernstein, I.M. Cleavage fracture in pearlitic eutectoid steel. *Metall. Trans. A* **1989**, *20*, 2321–2335. [CrossRef]
18. Echeverría, A.; Rodriguez-Ibabe, J.M. The role of grain size in brittle particle induced fracture of steels. *Mater. Sci. Eng. A* **2003**, *346*, 149–158. [CrossRef]
19. Curry, D.A.; Knott, J.F. Effects of microstructure on cleavage fracture stress in steel. *Met. Sci.* **1978**, *12*, 511–514. [CrossRef]
20. Linaza, M.A.; Rodriguez-Ibabe, J.M.; Urcola, J.J. Determination of the energetic parameters controlling cleavage fracture initiation in steels. *Fatigue Fract. Eng. Mater. Struct.* **1997**, *20*, 619–632. [CrossRef]
21. Balart, M.J.; Davis, C.L.; Strangwood, M.; Knott, J.F. Cleavage initiation in Ti-V-N and V-N microalloyed forging steels. *Mater. Sci. Forum* **2005**, *500–501*, 729–736. [CrossRef]
22. Du, J.; Strangwood, M.; Davis, C.L. Effect of TiN Particles and Grain Size on the Charpy Impact Transition Temperature in Steels. *J. Mater. Sci. Technol.* **2012**, *28*, 878–888. [CrossRef]
23. Rodriguez-Ibabe, J.M. The role of microstructure in toughness behaviour of microalloyed steels. *Mater. Sci. Forum* **1998**, *284*, 51–62. [CrossRef]
24. Bhadeshia, H. Physical metallurgy of steels. *Phys. Metall.* **2014**, *2014*, 2157–2214. [CrossRef]
25. Schindelin, J.; Arganda-Carreras, I.; Frise, E.; Kaynig, V.; Longair, M.; Pietzsch, T.; Preibisch, S.; Rueden, C.; Saalfeld, S.; Schmid, B.; et al. Fiji: An open-source platform for biological-image analysis. *Nat. Methods* **2012**, *9*, 676–682. [CrossRef] [PubMed]
26. Griffiths, J.R.; Owen, D.R.J. An elastic-plastic stress analysis for a notched bar in plane strain bending. *J. Mech. Phys. Solids* **1971**, *19*, 419–431. [CrossRef]
27. Bhadeshia, H.; Honeycombe, R. Chapter 13—Weld Microstructures. *Steels Microstruct. Prop.* **2017**, *2017*, 377–400. [CrossRef]
28. Shim, J.-H.; Oh, Y.-J.; Suh, J.-Y.; Cho, Y.W.; Shim, J.-D.; Byun, J.-S.; Lee, D.N. Ferrite nucleation potency of non-metallic inclusions in medium carbon steels. *Acta Mater.* **2001**, *49*, 2115–2122. [CrossRef]
29. Adrian, H. Microalloying '95. In Proceedings of the International Conference "Microalloying '95", ISS, Pittsburgh, PA, USA, 11–14 June 1995; p. 285.
30. Capdevila, C.; Caballero, F.G.; García-Mateo, C.; de Andrés, C.G. The role of inclusions and austenite grain size on intragranular nucleation of ferrite in medium carbon microalloyed steels. *Mater. Trans.* **2004**, *45*, 2678–2685. [CrossRef]
31. Sage, A.M. An overview of the use of microalloys in HSLA steels with particular reference to vanadium and titanium. *HSLA Steels Process. Prop. Appl.* **1990**, 51–60. [CrossRef]

32. Radović, N.; Koprivica, A.; Glišić, D.; Fadel, A.; Drobnjak, D. Influence of Cr, Mn and Mo on structure and properties of V microalloyed medium carbon forging steels. *MJoM* **2010**, *16*, 1053–1058.
33. Madariaga, I.; Gutiérrez, I. Nucleation of acicular ferrite enhanced by the precipitation of CuS on MnS particles. *Scr. Mater.* **1997**, *37*, 1185–1192. [CrossRef]
34. Sarma, D.S.; Karasev, A.V.; Jönsson, P.G. On the role of non-metallic inclusions in the nucleation of acicular ferrite in steels. *ISIJ Int.* **2009**, *49*, 1063–1074. [CrossRef]
35. Madariaga, I.; Romero, J.L.; Gutiérrez, I. Upper acicular ferrite formation in a medium-carbon microalloyed steel by isothermal transformation: Nucleation enhancement by CuS. *Metall. Mater. Trans. A* **1998**, *29*, 1003–1015. [CrossRef]
36. Jin, H.-H.; Shim, J.-H.; Cho, Y.W.; Lee, H.-C. Formation of Intragranular Acicular Ferrite Grains in a Ti-containing Low Carbon Steel. *ISIJ Int.* **2003**, *43*, 1111–1113. [CrossRef]
37. Zhang, S.; Hattori, N.; Enomoto, M.; Tarui, T. Ferrite Nucleation at Ceramic/Austenite Interfaces. *ISIJ Int.* **1996**, *36*, 1301–1309. [CrossRef]
38. Madariaga, I.; Gutierrez, I.; Bhadeshia, H. Acicular ferrite morphologies in a medium-carbon microalloyed steel. *Metall. Mater. Trans. A* **2001**, *32*, 2187–2197. [CrossRef]
39. Fadel, A.; Glišić, D.; Radović, N.; Drobnjak, D. Intragranular ferrite morphologies in medium carbon vanadium-microalloyed steel. *J. Min. Metall. Sect. B Metall.* **2013**, *49*, 237–244. [CrossRef]
40. Martin, J.I.S.; Rodriguez-Ibabe, J.M. Determination of energetic parameters controlling cleavage fracture in a Ti-V microalloyed ferrite-pearlite steel. *Scr. Mater.* **1999**, *40*, 459–464. [CrossRef]
41. Linaza, M.A.; Romero, J.L.; Rodriguez-Ibabe, J.M.; Urcola, J.J. Influence of the microstructure on the fracture toughness and fracture mechanisms of forging steels microalloyed with titanium with ferrite-pearlite structures. *Scr. Metall. Mater. States* **1993**, *29*. [CrossRef]
42. Gourgue, A.-F.; Flower, H.M.; Lindley, T.C. Electron backscattering diffraction study of acicular ferrite, bainite, and martensite steel microstructures. *Mater. Sci. Technol.* **2000**, *16*, 26–40. [CrossRef]
43. Linaza, M.A.; Romero, J.L.; Rodríguez-Ibabe, J.M.; Urcola, J.J. Cleavage fracture of microalloyed forging steels. *Scr. Metall. Mater.* **1995**, *32*, 395–400. [CrossRef]
44. Linaza, M.A.; Romero, J.L.; Rodriguez-Ibabe, J.M.; Urcola, J.J. Influence of the microstructure on the ductile-brittle behavior of engineering steels containing brittle particles. In Proceedings of the 36th Mechanical Working and Steel Processing Conference, Baltimore, MD, USA, 17–18 October 1994; pp. 483–494.
45. Smith, E. Cleavage fracture in mild steel. *Int. J. Fract. Mech.* **1968**, *4*, 131–145. [CrossRef]
46. Lindley, T.C.; Oates, G.; Richards, C.E. A critical of carbide cracking mechanisms in ferrite/carbide aggregates. *Acta Metall.* **1970**, *18*, 1127–1136. [CrossRef]
47. Ghosh, A.; Ray, A.; Chakrabarti, D.; Davis, C.L. Cleavage initiation in steel: Competition between large grains and large particles. *Mater. Sci. Eng. A* **2013**, *561*, 126–135. [CrossRef]
48. Lewandowski, J.J.; Thompson, A.W. Micromechanisms of cleavage fracture in fully pearlitic microstructures. *Acta Metall.* **1987**, *35*, 1453–1462. [CrossRef]
49. Strnadel, B.; Haušild, P. Statistical scatter in the fracture toughness and Charpy impact energy of pearlitic steel. *Mater. Sci. Eng. A* **2008**, *486*, 208–214. [CrossRef]
50. Glišić, D.; Fadel, A.; Radović, N.; Drobnjak, D.; Zrilić, M. Deformation behaviour of two continuously cooled vanadium microalloyed steels at liquid nitrogen temperature. *Hem. Ind.* **2013**, *67*, 981–988. [CrossRef]
51. Burns, K.W.; Pickering, F.B. Deformation+ fracture of Ferrite-pearlite Structures. *J. Iron Steel Inst.* **1964**, *202*, 899.
52. Irvine, K.J. Development of High Strength Steels. *J. Iron Steel Inst.* **1962**, *200*, 820.
53. Matlock, D.K.; Krauss, G.; Speer, J.G. Microstructures and properties of direct-cooled microalloy forging steels. *J. Mater. Process. Technol.* **2001**, *117*, 324–328. [CrossRef]
54. Caballero, F.G.; Garcia-Mateo, C. Phase transformations in advanced bainitic steels. In *Woodhead Publishing Series in Metals and Surface Engineering*; Pereloma, E., Edmonds, D.V., Eds.; Woodhead Publishing: Sawston, UK, 2012; Volume 2, pp. 271–294. [CrossRef]
55. Bhadeshia, H.K.D.H. *Bainite in Steels: Theory and Practice*, 3rd ed.; Maney Publishing: Leeds, UK, 2015.
56. Diaz, M.; Madariaga, I.; Rodriguez-Ibabe, J.M.; Gutierrez, I. Improvement of mechanical properties in structural steels by development of acicular ferrite microstructures. *J. Constr. Steel Res.* **1998**, *46*, 413–414. [CrossRef]
57. Echeverria, A.; Rodriguez-Ibabe, J.M. Cleavage micromechanisms on microalloyed steels. Evolution with temperature of some critical parameters. *Scr. Mater.* **2004**, *50*, 307–312. [CrossRef]

Article

Catastrophic Impact Loading Resilience of Welded Joints of High Strength Steel of Refineries' Piping Systems

Andrzej Klimpel [1], Anna Timofiejczuk [2], Jarosław Kaczmarczyk [3], Krzysztof Herbuś [4,*] and Massimiliano Pedot [2]

[1] The Department of Welding, The Faculty of Mechanical Engineering, The Silesian University of Technology, Konarskiego 18A, 44-100 Gliwice, Poland; andrzej.klimpel@polsl.pl

[2] The Department of Fundamentals of Machinery Design, The Faculty of Mechanical Engineering, The Silesian University of Technology, Konarskiego 18A, 44-100 Gliwice, Poland; anna.timofiejczuk@polsl.pl (A.T.); massimiliano.pedot@polsl.pl (M.P.)

[3] The Department of Theoretical and Applied Mechanics, The Faculty of Mechanical Engineering, The Silesian University of Technology, Konarskiego 18A, 44-100 Gliwice, Poland; jaroslaw.kaczmarczyk@polsl.pl

[4] The Department of Engineering Processes Automation and Integrated Manufacturing Systems, The Faculty of Mechanical Engineering, The Silesian University of Technology, Konarskiego 18A, 44-100 Gliwice, Poland

* Correspondence: krzysztof.herbus@polsl.pl

Citation: Klimpel, A.; Timofiejczuk, A.; Kaczmarczyk, J.; Herbuś, K.; Pedot, M. Catastrophic Impact Loading Resilience of Welded Joints of High Strength Steel of Refineries' Piping Systems. *Materials* **2022**, *15*, 1323. https://doi.org/10.3390/ma15041323

Academic Editors: Dražan Kozak, Nenad Gubeljak and Aleksandar Sedmak

Received: 1 December 2021
Accepted: 6 February 2022
Published: 11 February 2022

Publisher's Note: MDPI stays neutral with regard to jurisdictional claims in published maps and institutional affiliations.

Abstract: Refineries piping installation systems are designed, fabricated, and operated to assure very high levels of quality and structural integrity, to provide very high resilience to catastrophic events like earthquakes, explosions, or fires, which could induce catastrophic damage of piping systems due to collapse of nearby structures as towers, bridges, poles, walkways, vessels, etc. To evaluate the catastrophic impact loading resilience to failure of MMA (Manual Metal Arc Welding), GMA (Gas Metal Arc Welding), SSA (Self-shielded Arc Welding), and LASER+GMA of modern API 5L X80 pipes butt welded joints used for piping installation systems of refineries, the new, original technique of the quantitative and qualitative evaluation of impact loading resilience of butt welded joints of pipes was developed. The high-quality butt welded joints were impact loaded by the freely dropping 3000 kg mass hammer of the die forging hammer apparatus. The impact loading energy needed to exceed the yield strength of the extreme zone of welded joints and to induce catastrophic fracture of butt welded joints of API 5L X80 pipes was calculated using FEM (Finite Element Method) modeling of the impact loading process of tested butt welded joints of pipes. Results of the FEM modeling of impact loading technique of butt welded joints of piping systems indicate that it is a useful tool to provide valuable data for experimental impact loading tests of welded joints of pipes, decreasing the time and cost of the experiments. The developed impact loading technique of butt welded joints of pipes to simulate the catastrophic events in refinery piping systems and evaluate the resilience of the butt welded joints of pipes to catastrophic failure proved to be very efficient and accurate. Experiments of impact loading indicated that all specimens of butt welded joints API 5L X80 steel pipes are resilient to failure (cracks) in the extreme stressed/strained areas, above yield and tensile strength of the weld metals, no cracks or tears appeared in the extreme stressed/strained areas of the edges of the pipes, proving the very high quality of API 5L X80 steel pipes.

Keywords: refinery; piping; welded joint; API 5L X80 steel; nonlinear strength analyses; FEM

1. Introduction

Piping systems within refinery companies enable the continuous transfer of raw materials for the purposes of the assumed technological process of crude oil processing. Therefore, the condition of these systems directly impacts the safe operation of the company and ensures the required efficiency of the technological process. The piping systems of refineries include, among others, linear pipe sections, various types of pipe fittings, devices for forcing the circulation of the raw material (pumps), elements controlling its flow direction (valves), heat exchange systems, and systems for collecting the product. Welded joints are also considered the main parts of piping systems and as elements that are critical to the safe operation of refineries (Figure 1) [1–10]. The initial stage of the piping system design process is to define the functional requirements for the geometric form of its route. Meeting these assumptions will enable obtaining a safe transport route of the raw material from the starting point to the endpoint. On the other hand, its final form is influenced by such factors as the type of the transported raw material, the speed and the flow rate, the working pressure, the ambient temperature and the temperature of the transported fluid, results of the selection of design features based on strength analyses [4–6].

(a)

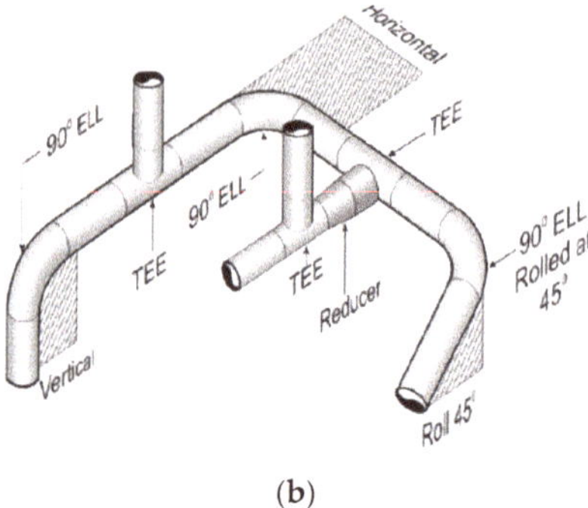

(b)

Figure 1. Example of the piping systems of refinery plants (**a**) A view of the magnitude of piping installations required in Jamnagar India refinery, the world's largest oil refinery with an aggregate capacity of 1.24 million barrels [7–9]; (**b**) A view of typical elements of the refinery piping installation systems where pipes butt joints are welded in horizontal–PC (2G) and vertical position–PH (5G) [7–9].

In order to determine safe values of the design features of the designed and manufactured pipe systems, with particular emphasis on welded joints, in addition to the previously mentioned values, the possibility of catastrophic events should be considered, such as the destruction of coexisting elements of the company's infrastructure. Items such as tanks, poles, or parts of a sidewalk falling onto the piping can be responsible for causing catastrophic impact loads. In addition, to ensure a very high level of resistance to catastrophic events, such as earthquakes, explosions, or fires, refinery piping systems are designed to provide a high level of integrity of the geometric form of their structure, with particular emphasis on preventing their unsealing [7,10]. The occurrence of a catastrophic situation causes an increase in the occurrence of dangerous situations for the company's staff, a break in the operation of the refinery infrastructure, and the necessity to carry out the process of diagnostics for damages and their removal. Critical areas of the piping system design process are related to the design of various types of connections, such as: flanged or welded. In relation to welded joints, it is very important to search for knowledge on the influence of possible impact loads on their properties. The data obtained in this way can be used to design specific structural nodes of future refinery piping systems.

One of the basic factors describing the failure of piping systems is the Pipe Diameter Factor (PDF). The PDF describes the relations between the pipe diameter and the possible severity of the failure. As the diameter of the pipe increases, the possible severity of failure increases. The PDF factor for a pipe of 12.0 inches (304 mm) in diameter is higher than for a pipe of 8.0 inches (203.2 mm). As the most modern and typical solution of piping systems, API 5L X80 high strength steel pipes 323.9 mm (12.75 inches) dia. and wall thickness 10.0 mm, HF longitudinally welded, produced by HUTA ŁABĘDY S.A. Poland were selected, and four welding processes were used to create specimens for catastrophic impact loading tests: MMA, GMA, SSA, and LASER + GMA. The research results presented in the paper are a continuation of research related to the quasi-static loading of the same welded joints of API 5L X80 steel pipes [11].

The following study was done to simulate and experimentally test the catastrophic impact loading resilience of the high-quality MMA, GMA, SSA, and LASER+GMA butt welded joints of API 5LX80 pipes of refineries piping systems:

- FEM modeling of the impact loading process of welded joints to establish the value of kinetic energy (the hammer mass and the hammer height of the forging hammer apparatus) of catastrophic load to provide the level of stresses and strains of the welded joint extreme stressed/strained areas, above the yield stress level of the parent material–API 5L X80 pipes and the weld metals to initiate cracks of the butt welded joints and parent material as well.
- Newly developed catastrophic impact loading tests of the butt welded joints specimens under impact energy calculated by FEM modeling to evaluate the resilience to impact loading and quality of welded joints of API 5L X80 pipes.

2. Preparation of Specimens of Butt Welded Joints of Pipes

The Welding Procedure Specifications (WPS) were worked out in Mostostal S.A. Zabrze, Poland, to prepare the butt welded joints specimens of sections of 120 mm width of API 5LX80 steel pipes 323.9 mm dia. and wall thickness 10.0 mm for catastrophic impact loading experiments (Table 1).

Table 1. Mechanical properties and chemical composition %wt. of API 5L X80 high strength steel pipes 323.9 mm dia. and wall thickness 10.0 mm, produced by HUTA ŁABĘDY S.A. [12].

$R_{0.5}$ MPa	R_m MPa	A5 %	KV at 0 °C J	C%	Si%	Mn%	Cr%	Ni%	Ti%	Al%	V%	Nb%
633	690	32	186–206	0.0766	0.232	1.33	0.2	0.019	0.0214	0.038	0.042	0.049

The welding conditions of three basic welding processes commonly used in the production of piping systems of refinery plants: MMA, GMA, and SSAW, and one, the most modern solution pipes' butt joints welding techniques–root pass laser welded and filling and cap passes—GMA welded, are presented in Table 2. The welding conditions of three basic welding processes commonly used in the production of piping systems of refinery plants: MMA, GMA, and SSAW, and one, the most modern solution pipes butt joints welding techniques–root pass laser welded and filling and cap passes—GMA welded, are shown in Table 2. The welding consumables (filer metals) were used for welded joints' specimens, assuring similar mechanical properties to the API 5L X80 high strength steel pipes, as shown in Tables 1 and 2.

Table 2. The welding conditions used to produce specimens of butt welded joints of API 5LX80 steel pipes 323.9 diameter and wall thickness 10.0 mm.

Welding Process/Welding Position	Filler Metal/ Polarity	Joint Prep.*	Passes	Filler Metal dia. [mm]	Welding Current [A]	Welding Voltage [V]	Travel Speed [cm/min]	Heat Input [kJ/cm]
MMA/PC-2G	Conarc 85 DC+	$\alpha = 50-60°$ b = 2−4 mm, c = 2.0 mm	root	2.5	65−75	21.0−23.0	5.0−7.0	9.4−16.6
			filling	3.2	115−125	23.0−26.0	16.0−24.0	5.5−9.8
			cap	3.2	115−125	23.0−26.0	16.0−24.0	5.5−9.8
MMA/PH-3G			root	2.5	65−75	21.0−23.0	3.0−5.0	13.1−27.6
			filling	3.2	110−120	23.0−26.0	12.0−16.0	7.6−12.5
			cap	3.2	110−120	23.0−26.0	12.0−16.0	7.6−12.5
GMA/PC-2G	LNM MoNiVa Shielding gas M21-flow rate = 12−16 L/min DC+	$\alpha = 50-60°$ b = 2−4 mm, c = 2.0 mm	root	1.2	100−110	16.0−19.0	10.0−13.0	5.9−10.0
			filling		190−220	21.5−23.5	28.0−34.0	5.8−8.9
			cap		190−220	21.5−23.5	28.0−34.0	5.8−8.9
GMA/PH-3G			root	1.2	90−105	16.0−18.0	9.0−12.0	5.8−10.1
			filling		160−180	18.0−20.0	22.0−28.0	5.0−12.3
			cap		160−180	18.0−20.0	12.0−16.0	8.6−14.4
SSA/PC-2G	PIPELINER NR-208XP DC-	$\alpha = 50-60°$ b = 2−4 mm, c = 2.0 mm	root	2.0	110−120	16.0−19.0	8.0−12.0	7.0−13.7
			filling		180−200	21.0−24.0	25.0−35.0	5.2−9.2
			cap		180−200	21.0−24.0	25.0−35.0	5.2−9.2
SSA/PH-3G			root	2.0	110−120	16.0−19.0	7.0−9.0	9.4−15.6
			filling		150−170	20.0−23.0	15.0−20.0	7.2−12.5
			cap		150−170	20.0−23.0	15.0−20.0	7.2−12.5
Root pas welding	Filler metal	Joint prep.	Beam quality and beam focusing system			Beam power [kW]	Welding speed [m/min]	Heat input [kJ/cm]
Laser Yb:YAG TruDisk 12002 fiber = 300 μm	No filler metal, shielding gas–Ar 12.0 L/min gas nozzle dia.= 8.0 mm	$\alpha = 60o$ b = 6.0 mm, c = 0.0 mm	$\leq$12.0 mm xmrad TRUMPF D70, fc = 200 mm, fcog = 400 mm, dcg = 0.8 mm			4.8	0.8	3.69

* Legend: *-α–bevel angle, b–root gap, c–root face. All joints preheat temperature–min 100 °C, interpass temperature-max. 250 °C.

3. Results of FEM Nonlinear Analysis of the Impact Loading Process of MMA, GMA, SSA, and LASER+GMA Butt Welded Joints of API 5LX80 Pipes

3.1. Material to Be Studied

The bilinear elastic-plastic material model of a welded joint of API 5L X80 steel pipes was adopted for performing the numerical simulations, as presented in Figure 2.

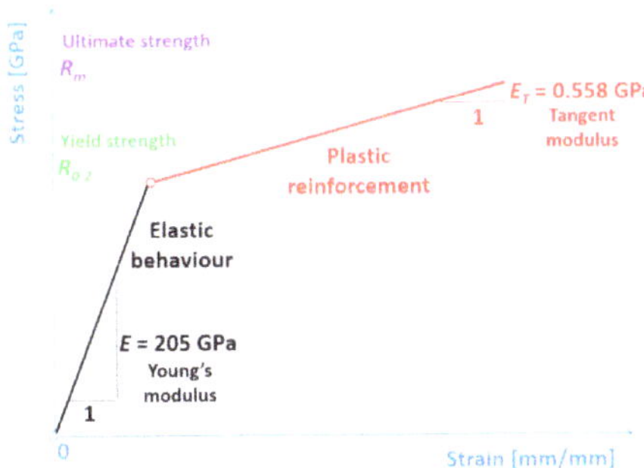

Figure 2. Bilinear elastic-plastic API 5L X80 steel pipes material model.

In Tables 1 and 3, the details of the weld metals and API 5L X80 steel pipes material properties for which the numerical calculations were performed are juxtaposed. The assumed bilinear material model belongs to the simplest but shows a nonlinear stress-strain behavior.

Table 3. Material properties of the weld metals and API 5LX80 steel pipes being impact loaded in the die forging hammer apparatus.

Basic Material Properties	Symbol	Value
Young's modulus	E	205 GPa
Poisson's ratio	ν	0.28
Kirchhoff's modulus	G	80 GPa
Tangent modulus	E_T	0.558 GPa
Yield strength	$R_{0.2}$	0.618 GPa
Ultimate tensile strength	R_m	0.700 GPa

3.2. Physical Model of the Butt Welded Joint of Pipes Specimens

The physical model of the MMA, GMA, SSA, LASER+GMA butt welded joints of sections of 120 mm width of API 5L X80 steel pipes, dia. 323.9 mm and wall thickness 10.0 mm, impact loaded in the die forging hammer apparatus is shown in Figure 3. The mean thickness of the welded joints reinforcement is 12.0 mm, and the mean width of the weld metal is 9.0 mm (the width of the weld face is approximately 16.0 mm, and the width of the weld root face is approximately 2.0 mm). Because the welded joints' HAZ (Heat Affected Zone) is very narrow, it was assumed to treat the HAZ as part of the weld metal. The butt welded joints of the pipes specimen have been divided into shell finite elements with five degrees of freedom in a node. Five Gauss integration points on the thickness of the shell of the butt welded joints of pipes were assumed. The weld metals and the pipes were mutually connected using the same nodes (without introduced contact) because they formed one inseparable entity.

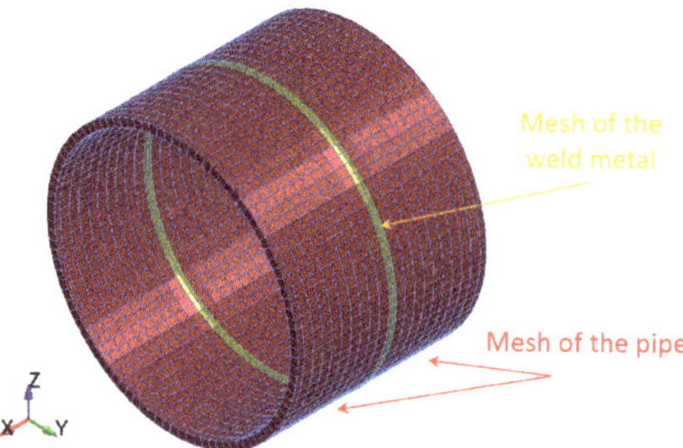

Figure 3. Discretization of the arc butt welded pipes into finite elements.

The specimens of butt welded joints of pipes supported by the base plate of the die forging hammer apparatus were impact-loaded by the hammer of the mass 3000 kg. Both hammer and base plate were modeled using solid elements with three degrees of freedom (Figure 4).

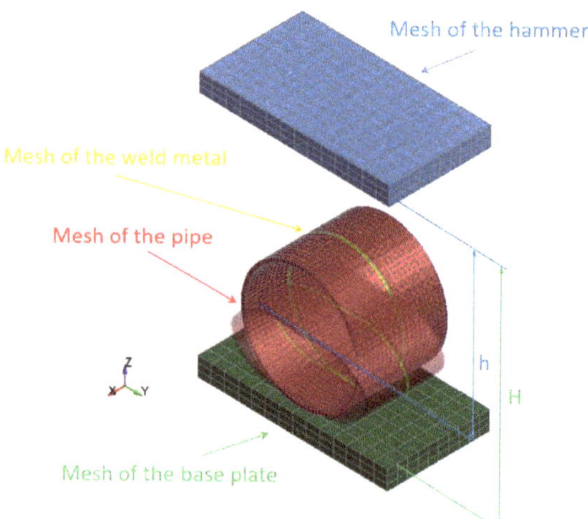

Figure 4. Discretization of the butt welded joint of pipes (a specimen), the hammer, and the base plate of the die forging hammer into finite elements.

When the specimen is placed on the base plate, the hammer is released from a certain height and free falls. Some preliminary cases were analyzed in the hammer height range 0.5 to 1.5 m, but finally, just two optimal cases were analyzed for the height: $H = 1.0$ m and $H = 1.5$ m. The distance between the initial hammer position and the top surface of the specimen to be impact loaded (marked as h) was introduced to estimate the potential energy which can be transformed into the plastic deformation of the specimen. The contact between the modeled welded joint of pipes (specimen) and the base plate as well as between the hammer and the specimen was introduced to avoid interpenetrating between the modeled parts (Figure 5).

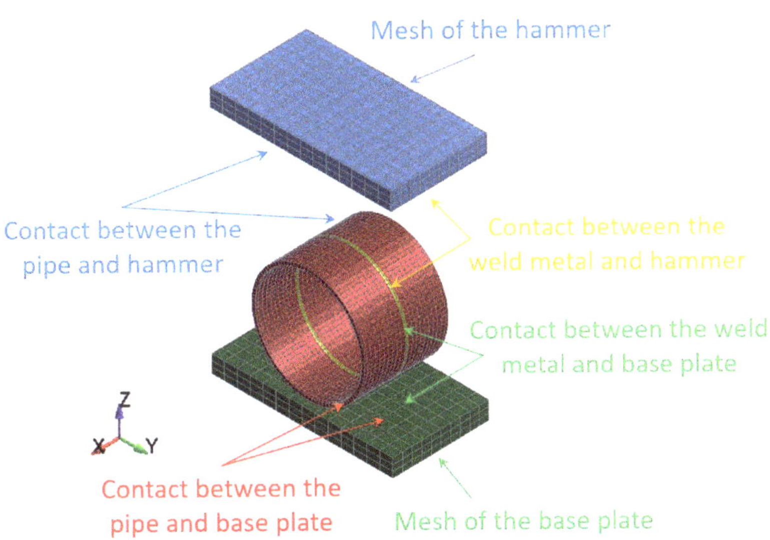

Figure 5. Contact details between modeled welded joint of pipes, the hammer, and the base plate.

Coulomb and Moren's model of friction was considered between all earlier mentioned contact surfaces. The static and kinetic coefficient of friction was assumed as for the steel. The static coefficient of friction equals ($\mu s = 0.15$), and the kinetic coefficient of friction was established ($\mu d = 0.1$), respectively, for all surfaces being in contact.

The detailed information concerning modeled impact-loaded parts created on the basis of elements and nodes are juxtaposed in Table 4. The most important parts are modeled as deformable (the weld metal and the pipe) because they are the subject of investigation. However, the hammer and base plate are modeled as rigid; therefore, their stiffness is much higher than the stiffness of the specimens of welded joints of pipes.

Table 4. The details concerning the individual parts of the physical model.

Parts	Type of Parts	Number of Elements	Number of Nodes
Pipe	Deformable	2000	2000
Hammer	Rigid	420	660
Base plate	Rigid	420	660
Weld metal	Deformable	200	300
Total in the model	-	3040	3620

3.3. Numerical Results

The impact loading process of the welded joint of API 5L X80 steel pipes was modeled using the finite element method–FEM and computer system LS-DYNA. The numerical calculations results were juxtaposed for several successive time intervals to facilitate the investigation of the mechanism of the process using the impact loading hammer and the base plate of the die forging hammer apparatus. Two optimal variants were researched for two different heights: H = 1.0 m and H = 1.5 m, respectively, as shown in Figures 6 and 7.

To facilitate the analysis of the data shown in Figures 6 and 7, the obtained results were juxtaposed in two tables for two cases for H = 1.0 m and H = 1.5 m, respectively (Tables 5 and 6). The time and corresponding Huber–Mises stress and effective plastic strain are presented in these tables.

Table 5. The juxtaposition of Huber–Mises stresses and effective plastic strains during impact loading of the specimens of welded joints of pipes for height H = 1.0 m, Figure 6.

No	Time [ms]	Huber–Mises Stress [MPa]	Effective Plastic Strain [mm/mm]
1.	373	403	0.000
2.	374	618	0.0019
3.	384	651	0.0587
4.	394	655	0.0668
5.	404	680	0.1110
6.	414	703	0.1519
7.	424	724	0.1908
8.	434	742	0.2225
9.	444	763	0.2597
10.	454	779	0.2873
11.	464	794	0.3155
12.	474	800	0.3256
13.	480	782	0.3260
14.	481	762	0.3260

Table 6. The juxtaposition of Huber–Mises stresses and effective plastic strains during impact loading of the specimens of welded joints of pipes for height H = 1.5 m.

No	Time [ms]	Huber–Mises Stress [MPa]	Effective Plastic Strain [mm/mm]
1.	491	426	0.0000
2.	492	619	0.0029
3.	502	647	0.0604
4.	512	676	0.1031
5.	522	708	0.1616
6.	532	743	0.2236
7.	542	785	0.2989
8.	552	835	0.3888
9.	562	894	0.4937
10.	572	940	0.5754
11.	578	760	0.5951
12.	579	623	0.5951

To check the correctness of numerical calculations, the energy balance of the welded joint of pipes (specimen) was determined for two analyzed cases, H = 1.0 m and H = 1.5 m, respectively (Figures 8 and 9). The initial potential energy could be compared with the total energy for the final time instant.

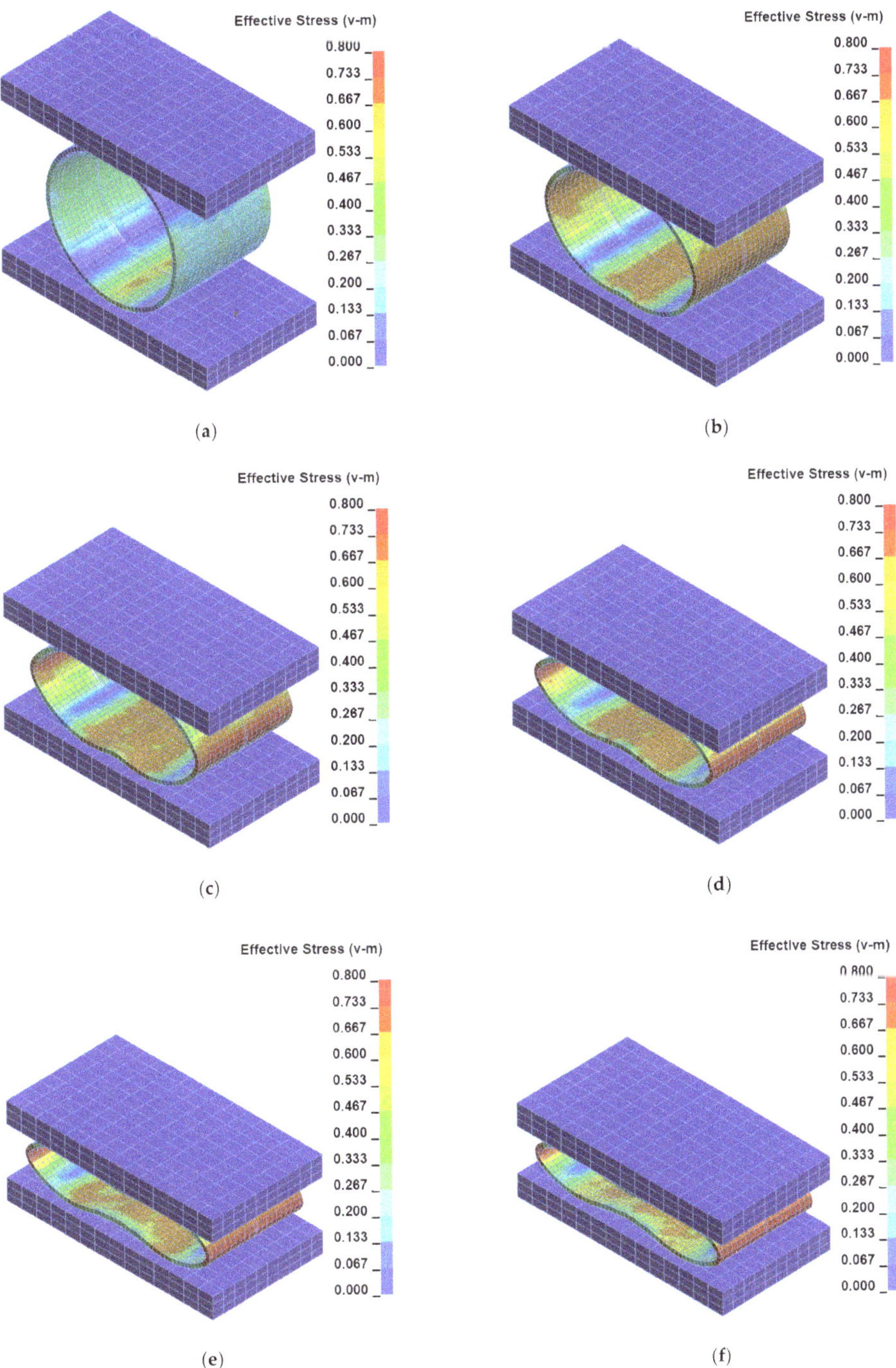

Figure 6. Distribution of the equivalent Huber–Mises stresses [GPa] for arbitrary selected time instants [ms]: (**a**) t = 374, (**b**) t = 394, (**c**) t = 414, (**d**) t = 434, (**e**) t = 454, (**f**) t = 481 for height H = 1.0 m, Table 5.

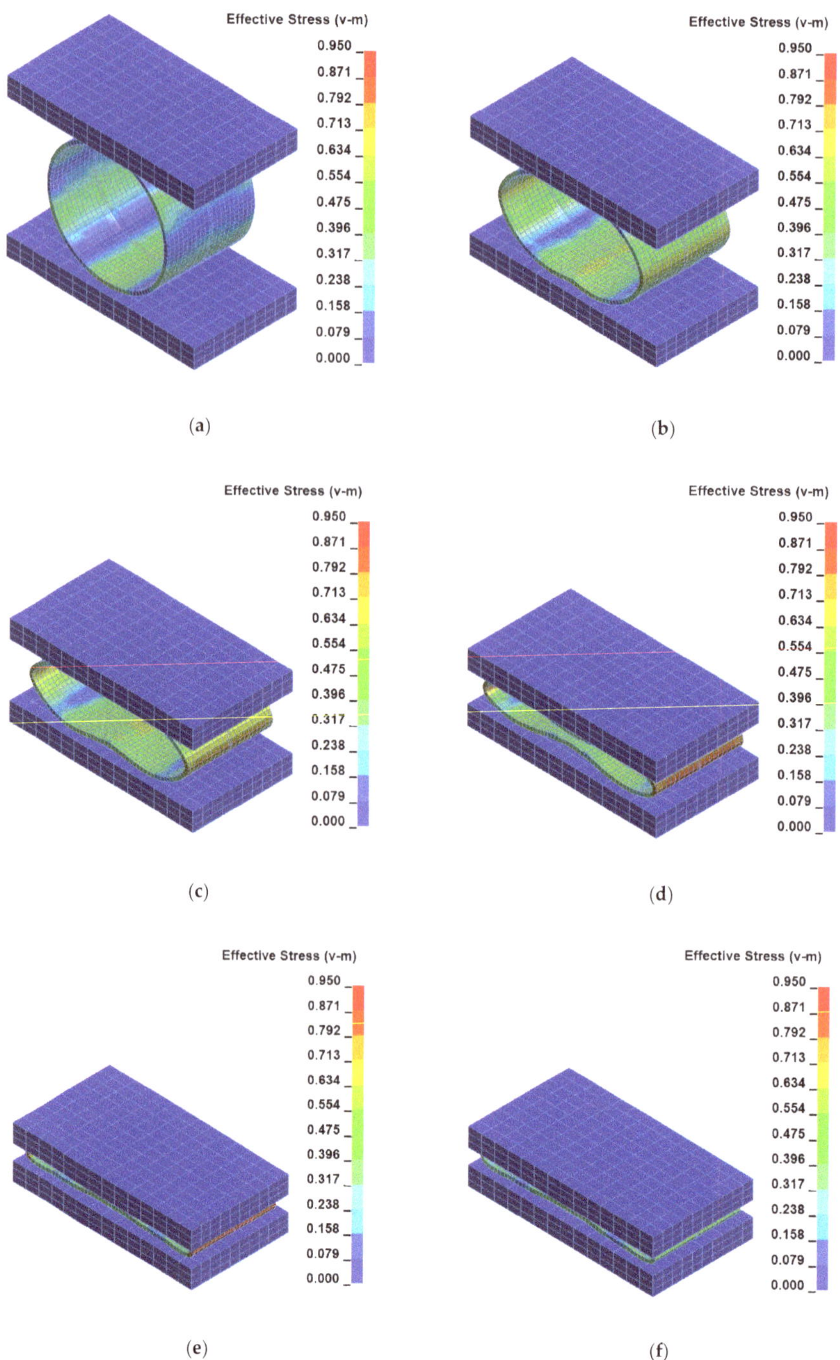

Figure 7. Distribution of the equivalent Huber–Mises stresses [GPa] for arbitrary selected time instants [ms]: (**a**) t = 492, (**b**) t = 512, (**c**) t = 532, (**d**) t = 552, (**e**) t = 572, (**f**) t = 579 for height H = 1.5 m, Table 6.

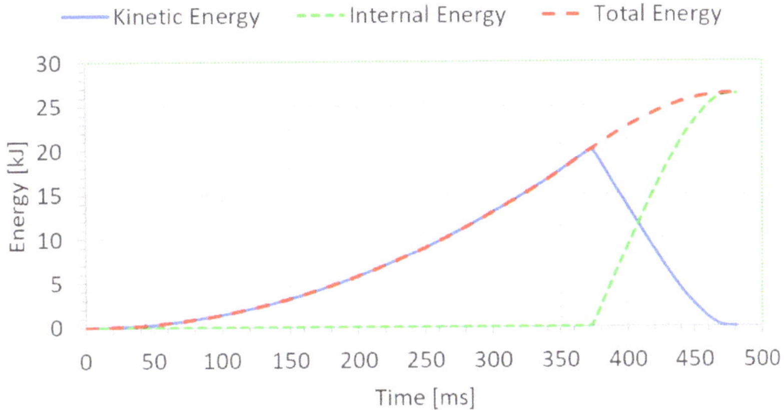

Figure 8. The energy balance [kJ] of the welded joint of pipes for height H = 1.0 m.

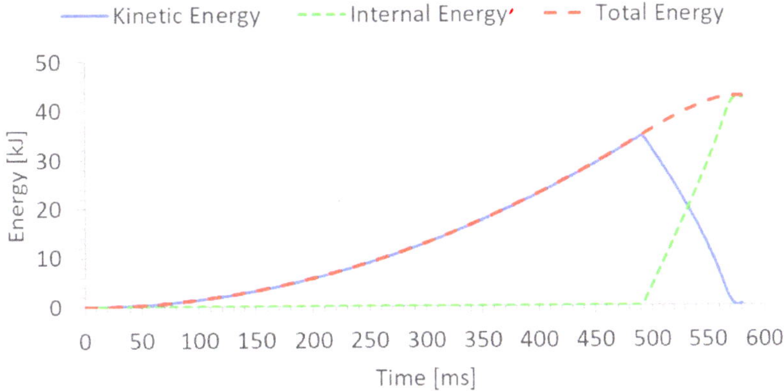

Figure 9. The energy balance [kJ] of the welded pipe for height H = 1.5 m.

The potential energy can be calculated according to the following formula:

$$E_p = m \cdot g \cdot h, \tag{1}$$

where:
m—mass of the hammer [kg],
g—gravitational acceleration [m/s^2],
h—height measured from the initial position of the hammer to the final position of the top surface of the deformed welded joint of pipes [m].

The mass of the hammer is 3000 kg. The distance between the initial hammer position and the top surface of the deformed welded joint of pipes (marked as h) can be found based on the graph (Figure 10) and estimated using, for example, the following formula:

$$h = H + h_1 - D, \tag{2}$$

where:
H—height measured from the initial position of the hammer and the top surface of the base plate [m],
D—external diameter of the welded joint of pipes [m],

h_1—displacement of the welded joint of pipes [m] obtained from the graph for the green color (Figure 10); it is given in millimeters, and it should be recalculated into meters.

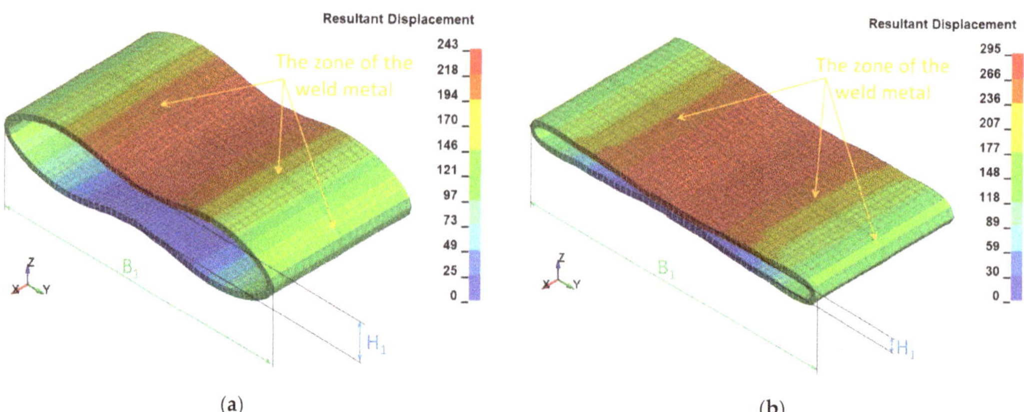

Figure 10. The resultant displacements [mm] in the welded joint of pipes for two variants: (**a**) height H = 1.0 m (H1 = 64 mm; B1 = 461 mm), (**b**) height H = 1.5 m (H1 = 20 mm; B1 = 481 mm).

The potential energy calculated in this manner from Equation (1) can serve to compare it with the total energy or internal energy for the final time instant shown on the graph (Figures 8 and 9) depending on the variant of height (H = 1.0 and H = 1.5 m, respectively).

The following resultant values of several selected physical quantities such as displacements (Figure 10), Huber–Mises stresses (Figure 11), effective plastic strains (Figure 12), as well as thickness of the welded pipe (Figure 13) for the final position of the deformed welded joint of pipes are juxtaposed consecutively for two considered variants of height H = 1.0 m and H = 1.5 m, respectively. These data are intended to initially estimate the parameters of the actual experiment.

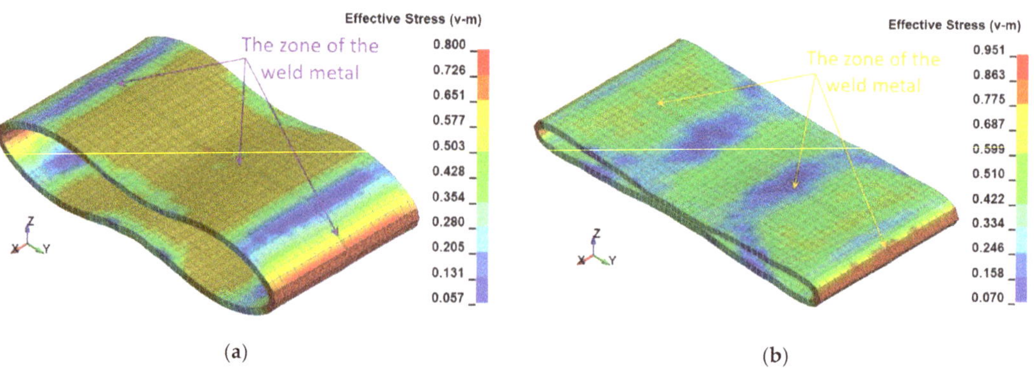

Figure 11. The Huber–Mises stresses [GPa] in the welded joint of pipes for two variants: (**a**) height H = 1.0 m, (**b**) height H = 1.5 m.

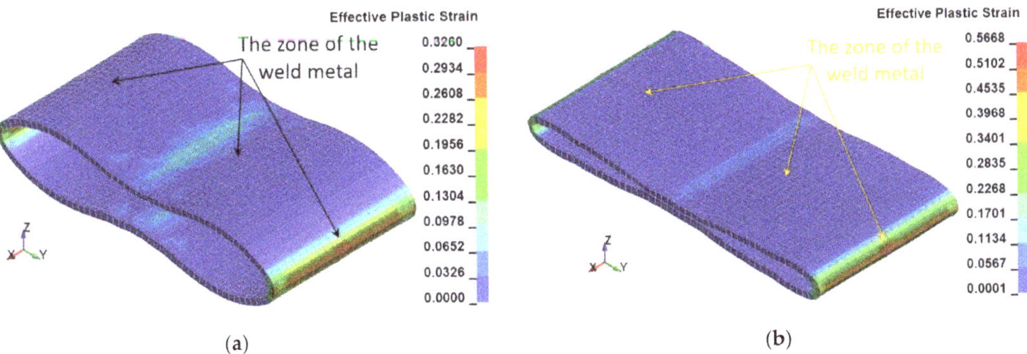

Figure 12. The Huber–Mises effective plastic strains [mm/mm] in the welded joint of pipes for two variants: (**a**) height H = 1.0 m, (**b**) height H = 1.5 m.

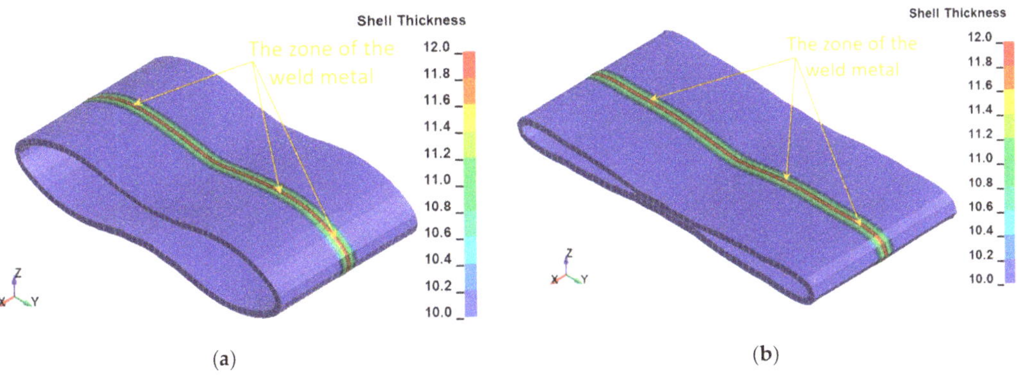

Figure 13. The thickness [mm] of the welded joint of pipes for two variants: (**a**) height H = 1.0 m, (**b**) height H = 1.5 m.

The FEM modeling analysis indicated that the resultant values for the Huber–Mises Stresses, effective plastic strains, and displacements are much higher in the case of larger height H = 1.5 m, Figures 10–12. However, the thickness of the wall of the welded joint of pipes changes insignificantly, and in consequence, this variation can be neglected, as shown in Figure 13. It was estimated that the impact load of the specimen of welded joint of pipes at the hammer height H = 1.0 m induced Huber–Mises stress equal to 800 MPa and effective plastic strain of 0.3256 mm/mm. At the hammer height H = 1.5 m, the impact load-induced Huber–Mises stress equal to 940 MPa and effective plastic strain of 0. 5754 mm/mm. In both cases, the Huber–Mises stress was much higher than the yield strength and tensile strength of MMA, GMA, SSA, and LASER+GMA welded joints of API 5L X800 steel pipes weld metals and parent material, Tables 3, 5 and 6, Figures 6–12.

4. Impact Loading Experiments of MMA, GMA, SSA, and LASER+GMA Butt Welded Joints of API 5L X80 Pipes Specimens

The developed impact loading technique to test the resilience to catastrophic impact loading of the MMA, GMA, SSA, and LASER+GMA butt welded joints of API 5L X80 steel pipes was executed at KUZNIA ŁABĘDY S.A. plant (www.kuznia-labendy.pl, accessed on 25 November 2021) on the die forging hammer SKM-3T apparatus (Figure 14), at the load of the 3000 kg weight (mass) of the freely dropping hammer and two hammer heights 1.0 and 1.5 m, selected on the bases of the results of analysis of FEM modeling of impact loading of welded joint of pipes, Tables 5 and 6, Figures 6–12. The scheme of impact loading tests of butt welded joints of API 5L X80 steel pipes is shown in Figure 15. The first impact loading test was done for the MMA butt welded joint specimen at the hammer load of 3000 kg from the height of 1.0 m. The impact loaded specimens were flattened to the geometrical dimensions H and B [mm], and no cracks in extreme stressed/strained areas of the butt welded joint or pipes edges were detected, despite the level of FEM calculated Huber–Mises stresses were over tensile strength of the weld metal and the API 5L X80 steel (Figure 16).

(a)

(b)

Figure 14. A view of (**a**) the die forging hammer 3000 kg mass, Russian production SKM-3 T apparatus, (**b**) the specimen of MMA butt welded joint of API 5L X80 steel pipes before impact loading test.

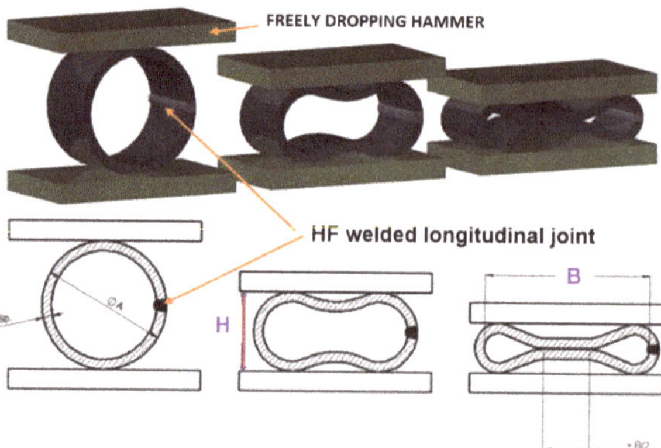

Figure 15. The scheme of impact loading test of butt welded joints of pipes: H and B–geometrical parameters of the deformation [13,14].

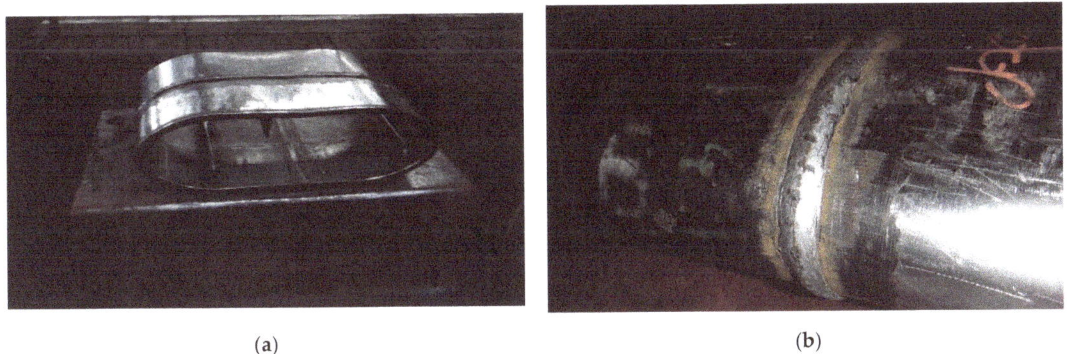

Figure 16. A view of MMA butt welded joint of API 5L X80 steel pipes after the 3000 kg impact loading test–the hammer height 1.0 m. No cracks of both extreme stressed/strained areas of the MMA butt welded joint (**a**,**b**).

To force failure (cracks) in extreme stressed/strained areas of the MMA butt welded joint, impact loading energy was increased by 25%, and the 3000 kg hammer height was increased to 1.5 m. After this second impact load test, the specimen was flattened to the plate shape, and both extremes stressed/strained areas of the MMA butt welded joint specimen cracked, but surprisingly no cracks appeared on the MMA welded pipes edges (Figures 17–19). Results of impact loading test at 3000 kg hammer mass and the 1.0 m hammer height of GMA, SSA, and LASER+GMA butt welded joints of API 5L X80 pipes are shown in Figures 19–22 and Table 7.

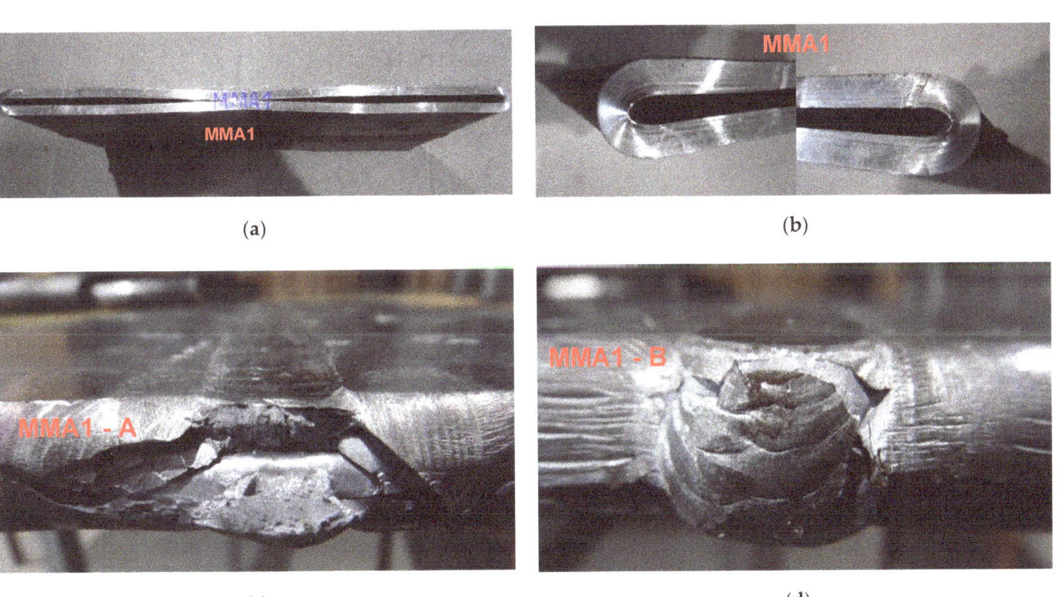

Figure 17. A view of cracked both extreme areas of stressed/strained A and B of MM1 butt welded joint of API 5L X80 steel pipes specimen and pipes edges after second the 3000 kg impact loading test of the specimen of MMA welded joint (the hammer height 1.5 m)–(**c**,**d**). No cracks or tears of the extreme stressed/strained areas of the edges of the pipes (**a**,**b**).

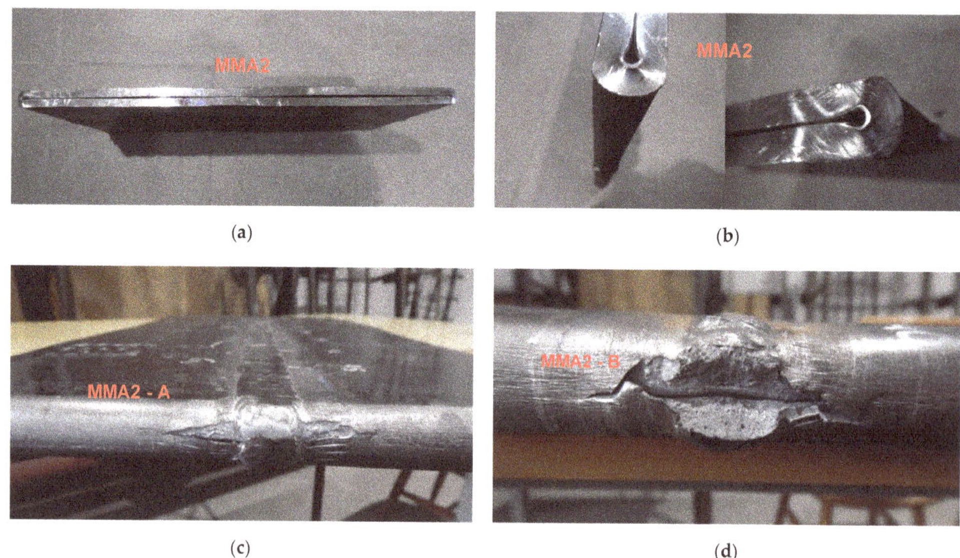

(a)

(b)

(c)

(d)

Figure 18. A view of cracked both extreme stressed/strained areas of MM2-A–(**c**) and MM2-B (**d**), of welded joint of API 5L X80 steel pipes specimen and pipes edges after the 3000 kg impact loading test (the hammer height 1.5 m). No cracks or tears of the extreme stressed/strained areas of the edges of the pipes–(**a**,**b**).

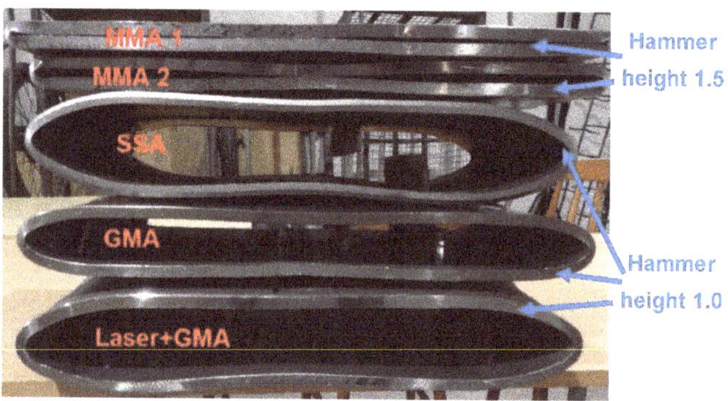

Figure 19. A view of MMA1, MMA2, GMA, SSA, and LASER+GMA butt welded joints of API 5L X80 steel pipes specimens after the 3000 kg hammer impact loading tests (the hammer height 1.0 m and 1.5 m).

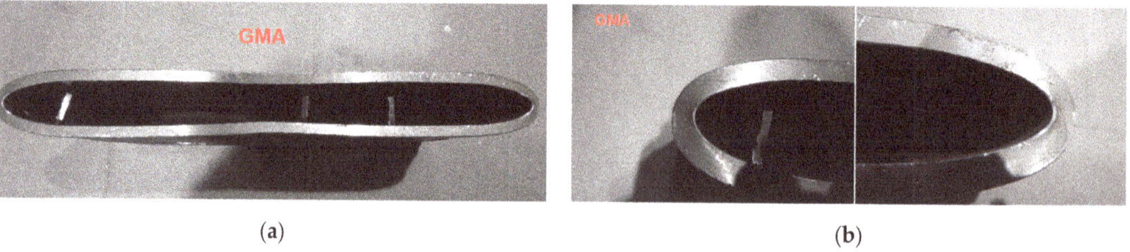

(a)

(b)

Figure 20. *Cont.*

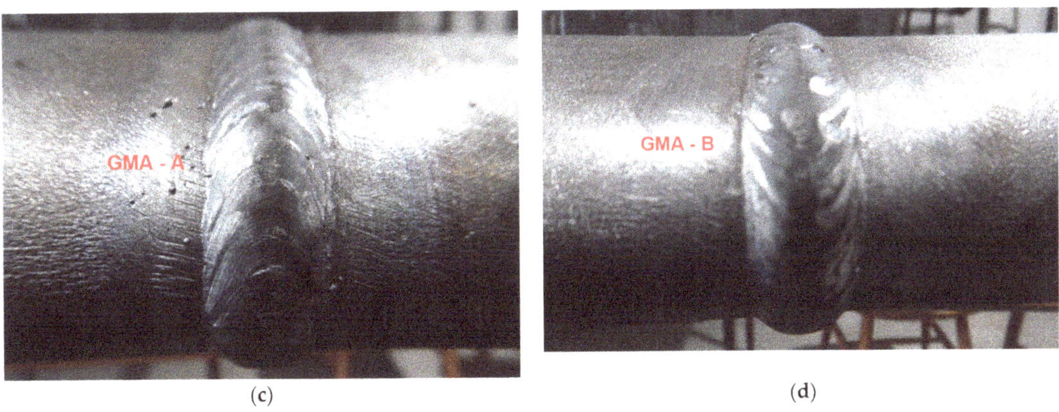

Figure 20. A view of GMA butt welded joint of API 5L X80 steel pipes and pipes edges after the 3000 kg impact loading test (the hammer height 1.0 m). No cracks of both extreme stressed/strained areas of the butt welded joint (**c,d**), and no cracks and tears of the extreme stressed/strained areas of the edges of the pipes (**a,b**).

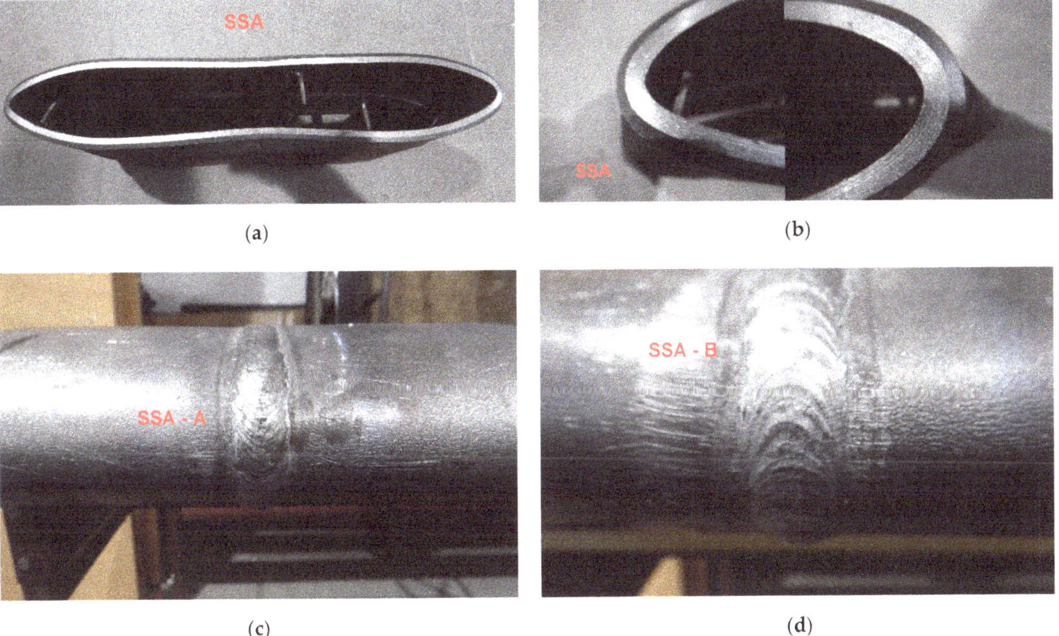

Figure 21. A view of SSA butt welded joint of API 5L X80 steel pipes and pipes edges after the 3000 kg impact loading test (the hammer height 1.0 m). No cracks of both extreme stressed/strained areas of the butt welded joint (**c,d**), and no cracks and tears of the extreme stressed/strained areas of the edges of the pipes (**a,b**).

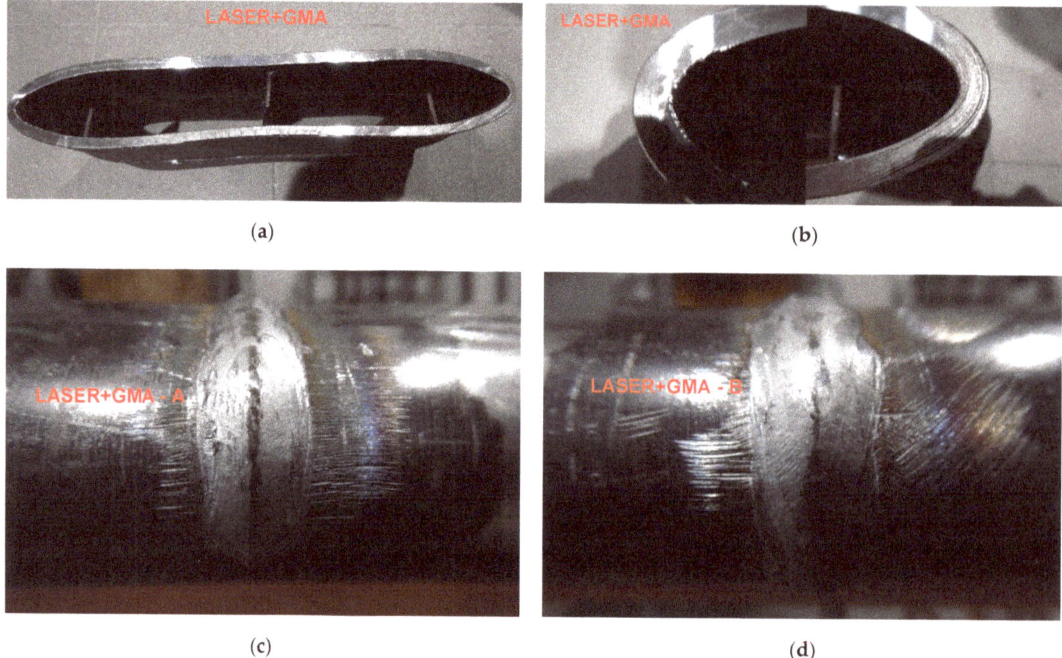

(a)

(b)

(c)

(d)

Figure 22. A view of LASER+GMA butt welded joint of API 5L X80 steel pipes and pipes edges after the 3000 kg impact loading test (the hammer height 1.0 m). No cracks of both extreme stressed/strained areas of the butt welded joint (**c,d**), and no cracks and tears of the extreme stressed/strained areas of the edges of the pipes (**a,b**).

Table 7. Results of impact loading tests of MMA, GMA, SSA, and LASER+GMA butt welded joints of API 5L X80 pipes specimens.

Type of Joint	H [mm]	B [mm]	Quality of Welded Joint	Quality of Pipes Edges
MMA	74	456	no cracks	no cracks
MMA1	20	485	cracks	no cracks
MMA2	20	488	cracks	no cracks
GMA	75	466	no cracks	no cracks
SSA	78	463	no cracks	no cracks
LASER+GMA	55	474	no cracks	no cracks

5. Conclusions

The FEM nonlinear analysis of the MMA, GMA, SSA, and LASER+GMA butt welded joints of API 5L X80 steel pipes, 323.9 mm dia. and wall thickness 10.0 mm of the resilience to catastrophic impact loading of welded joints of pipes by the hammer of the die forging hammer apparatus, Figure 14, indicates as follows:

1. The numerical simulations of the catastrophic impact loading of butt welded joints of pipes' specimen have been performed by using the FEM modeling and the computer system LS-DYNA. The explicit analysis has been carried out, considering the pipes' material properties and geometrical nonlinearities. Results of numerical simulations indicate that the weld metal + HAZ does not crack and also parent material or it is not submitted to any failure even though the Huber–Mises stresses and effective plastic strains are beyond the yield strength ($R_e = 0.618$ GPa) or even tensile strength ($R_m = 0.700$ GPa) of the weld

metal and API 5L X80 steel parent material, Tables 3, 5 and 6, Figures 6–12. The dynamic fast-changing numerical simulation shows that the Huber–Mises stresses reach such values as 0.8 GPa for the impact loading hammer height H = 1.0 m and 0.95 GPa for the hammer height H = 1.5 m. The energy balance has been conducted to confirm the correctness of the obtained numerical results. The potential energy (Ep = m·g·h) corresponds approximately to the total energy for the final position of the butt welded joints of API 5L X80 steel pipes (after deformation) obtained from the presented graphs (Figures 8 and 9) for both analyzed variants of the impact loading hammer height (H = 1.0 m and 1.5 m). From the data obtained using numerical calculations concerning the selected physical quantities, the following conclusions can be drawn:

- Along with the increase of the height of the released hammer:
  - ✓ the equivalent Huber–Mises stresses grow;
  - ✓ the equivalent effective plastic strains grow;
  - ✓ the resultant displacements grow;
- Along with the increase of the height of the impact loading hammer, the thickness of the arc butt welded joints and wall thickness of pipes changes insignificantly which could be neglected.

Finally, the height and width of the welded joints of pipes' specimens after deformation obtained from the numerical calculations were compared with data received from experiments of hammer impact loading. The comparison of numerical and experimental results demonstrates a good agreement, proving that FEM simulation of technological processes is a useful tool to support experimental study.

2. The developed impact loading technique to evaluate the resilience to catastrophic failure of the MMA, GMA, SSA, and LASER+GMA butt welded joints of API 5L X80 steel pipes under simulated catastrophic impact loading events in refineries piping systems and proved to be very efficient and accurate. All impact loaded specimens of the MMA, GMA, SSA, and LASER+GMA butt welded joints of API 5L X80 steel pipes at impact energy forced by the hammer mass 3000 kg and at the hammer height H = 1.0 m, forcing at the extreme areas of butt welded joints the Huber–Mises stresses and effective plastic strains beyond the yield strength (R_e = 0.618 GPa) or even tensile strength (R_m = 0.700 GPa) of the weld metal and API 5L X80 steel pipes, proved high resilience to catastrophic impact loads, as no cracks or tears were detected, Tables 1, 5 and 7, Figures 16 and 19, Figures 20–22.

3. Impact loaded specimens of MMA butt welded joints of API 5L X80 steel pipes at impact energy forced by the hammer mass 3000 kg and at the hammer height H = 1.5 m, forcing stresses and strains at the extreme areas of MMA butt welded joints and the parent material of pipes approximately 25% higher than the tensile strength of weld metals and parent material of pipes, Tables 1 and 6, resulted in total flattening of the MMA butt welded joints of pipes specimens to H = 20 mm (double thickness of the pipes t = 10 mm). All welded joints strongly cracked at the extreme areas, but no cracks or tears appeared on the extreme edges of pipes, proving the very high quality of API 5L X80 steel pipes tested, Table 7, Figures 17, 18 and 22.

Author Contributions: Conceptualization, A.K. and A.T.; methodology, A.K.; software, J.K. and K.H.; validation, A.K., J.K. and K.H.; formal analysis, A.T. and M.P.; investigation, M.P.; resources, A.K. and A.T.; data curation, A.K. and A.T.; writing—original draft preparation, A.K., J.K. and K.H.; writing—review and editing, A.K., A.T. and K.H.; visualization, M.P.; supervision, A.T.; project administration, A.K. and M.P.; funding acquisition, A.K. All authors have read and agreed to the published version of the manuscript.

Funding: The paper is the result of works carried out the project titled: Extreme loading analysis of petrochemical plants and design of metamaterial-based shields for enhanced resilience (XP-RESILIENCE), funding under the European Union's Horizon 2020—H2020-EU.1.3.1. (Project Reference: 721816, Call: H2020-MSCA-ITN-2016, Period: September 2016–August 2020).

Institutional Review Board Statement: Not applicable.

Informed Consent Statement: Not applicable.

Data Availability Statement: Not applicable.

Conflicts of Interest: The authors declare no conflict of interest.

References

1. Paolicci, F.; Reza, S.; Bursi, O. Rfcs—Induse Project: Structural Safety of Industrial Steel Tanks, Pressure Vessels and Piping Systems Under Seismic Loading, Grant No. RFSR-CT2009-00022. Available online: https://op.europa.eu/en/publication-detail/-/publication/dfb8b89c-8de0-435c-b262-9acec59f318d (accessed on 25 November 2021).
2. Vicente, F. Criticality assessment of piping systems for oil & gas facilities. *Insp. J.* **2014**, *20*, 15–18.
3. Rintamaa, R. *Prevention QF Catastrophic Failure in Pressure Vessels and Pipings*; Final Report of the NKA-Project MAT 570. Nordic Liaison Committee for Atomic energy; International Atomic Energy Agency (IAEA): Vienna, Austria, 1989.
4. Simonen, F.A. *Pressure Vessels and Piping Systems: Reliability, Risk and Safety Assessment*; Ancillary Equipment and Electrical Equipment; Pacific Northwest National Laboratory: Richland, WA, USA, 2010; Available online: http://www.desware.net/sample-chapters/d09/e6-165-07-00.pdf (accessed on 25 November 2021).
5. Brickstad, B.; Schimpfke, T. *Benchmarking of Structural Reliability Models for Risk Analyses of Piping*; Swedish Nuclear Power Inspectorate (SKI): Stockholm, Sweden; Gesellschaft für Anlagen und Reaktorsicherheit (GRS) mbH: Köln, Germany, 2005; Available online: https://pdfs.semanticscholar.org/dc95/7098e3047f8fb6c030fc3a9df0fceeef4fef.pdf?_ga=2.142886710.2063904234.1598616093-209817455.1598616093 (accessed on 25 November 2021).
6. *CSB Investigations Involving Inadequate Mechanical Integrity Programs*; Chevron Refinery Fire; Chevron U.S.A. Inc.: Richmond, CA, USA, 2005. Available online: https://www.csb.gov/recommendations/csb-investigations-mechanical-integrity/ (accessed on 25 November 2021).
7. Abduh, M. Erosion Corrosion—Learning from Humber Estuary. 2009. Available online: https://abduh137.wordpress.com/category/engineering-failures/https://abduh137.worldpress.com/category/engineering-failures/ (accessed on 25 November 2021).
8. Available online: https://www.statista.com/statistics/981799/largest-oil-refineries-worldwide/ (accessed on 25 November 2021).
9. Available online: https://www.listnbest.com/11-worlds-largest-oil-refineries-processing-capacity/ (accessed on 25 November 2021).
10. Available online: https://www.google.com/search?sxsrf=ALeKk03vC_L27-bIDDawgj491KzcLl1Zfg:1589638579626&source=univ&tbm=isch&q=refinery+piping+systems+catastrophic+failures&client=firefox-b-d&sa=X&ved=2ahUKEwikgfCWybjpAhV3DGMBHebQCPsQsAR6BAgEEAE&biw=1231&bih=707 (accessed on 25 November 2021).
11. Klimpel, A.; Herbuś, K.; Ociepka, P.; Timofiejczuk, A.; Pedot, M. *Quasi-Static Loading of Piping Welded Joints*; International Journal of Modern Manufacturing Technologies: Iași, Romania, 2020; Available online: https://www.ijmmt.ro/vol12no22020/10_Andrzej_Klimpel_1.pdf (accessed on 25 November 2021).
12. *Mill Test Certificate NO 160818*; POSCO. HUTA ŁABĘDY S.A: Gliwice, Poland, 2019.
13. *EN ISO 8492:2004*; Metallic Materials—Tube—Flattening Test. ISO: London, UK, 2004.
14. *ASTM 370*; Standard Test Methods and Definitions for Mechanical Testing of Steel Products. ASTM: West Conshohocken, PA, USA, 2019.

 *materials*

Article

Contact Reactive Brazing of TC4 Alloy to Al7075 Alloy with Deposited Cu Interlayer

Mengjuan Yang [1], Chaonan Niu [2], Shengpeng Hu [2], Xiaoguo Song [2,3,*], Yinyin Pei [4], Jian Zhao [1] and Weimin Long [4]

1   School of Materials Engineering, Shanghai University of Engineering Science, Shanghai 201620, China; yang13137760150@163.com (M.Y.); zhaojianhit@163.com (J.Z.)
2   State Key Laboratory of Advanced Welding and Joining, Harbin Institute of Technology, Harbin 150001, China; ncnhitwh@163.com (C.N.); sp_hu@hit.edu.cn (S.H.)
3   Shandong Institute of Shipbuilding Technology, Weihai 264209, China
4   State Key Laboratory of Advanced Brazing Filler Metals and Technology, Zhengzhou Research Institute of Mechanical Engineering Co., Ltd., Zhengzhou 450001, China; balloy@163.com (Y.P.); brazelong@163.com (W.L.)
*   Correspondence: xgsong@hitwh.edu.cn; Tel./Fax: +86-631-5678-454

Abstract: The brazing of Titanium alloy to Aluminum alloy is of great significance for lightweight application, but the stable surface oxide film limits it. In our work, the surface oxide film was removed by the ion bombardment, the deposited Cu layer by magnetron sputtering was selected as an interlayer, and then the contact reactive brazing of TC4 alloy to Al7075 alloy was realized. The microstructure and joining properties of TC4/Al7075 joints obtained under different parameters were observed and tested, respectively. The results revealed that the intermetallic compounds in the brazing seam reduced with the increased brazing parameters, while the reaction layer adjacent to TC4 alloy continuously thickened. The shear strength improved first and then decreased with the changing of brazing parameters, and the maximum shear strength of ~201.45 ± 4.40 MPa was obtained at 600 °C for 30 min. The fracture path of TC4/Al7075 joints changed from brittle fracture to transgranular fracture, and the intergranular fracture occurred when the brazing temperature was higher than 600 °C and the holding time exceeded 30 min. Our work provides theoretical and technological analyses for brazing TC4/Al7075 and shows potential applications for large-area brazing of titanium/aluminum.

Keywords: Al7075; TC4; contact reactive brazing; Cu deposited

Citation: Yang, M.; Niu, C.; Hu, S.; Song, X.; Pei, Y.; Zhao, J.; Long, W. Contact Reactive Brazing of TC4 Alloy to Al7075 Alloy with Deposited Cu Interlayer. *Materials* **2021**, *14*, 6570. https://doi.org/10.3390/ma14216570

Academic Editors: Nenad Gubeljak, Dražan Kozak and Aleksandar Sedmak

Received: 17 September 2021
Accepted: 27 October 2021
Published: 1 November 2021

Publisher's Note: MDPI stays neutral with regard to jurisdictional claims in published maps and institutional affiliations.

1. Introduction

Ti-6Al-4V (TC4) alloy has unique properties such as low thermal conductivity, superior corrosion resistance, superior mechanical properties, high-temperature strength, and low-temperature toughness, which has attracted wide attention in the aerospace field [1–4]. Al7075 alloy, which has low density, high specific strength, casting properties, good corrosion resistance, and high conductivity, has been widely used in structural parts of the aerospace field [5–7]. Currently, TC4 and Al7075 alloys are simultaneously used in composite components of aircraft wings and automotive airfoils, where the performance of the components can be improved by combining the advantages of the two materials [8–12].

At present, the methods of joining Al alloys to TC4 alloys mainly include laser welding [13,14], transient liquid phase (TLP) bonding [15], brazing [16], diffusion bonding [17,18], etc. Among them, brazing is suitable for joining dissimilar materials with a large difference in physical and chemical properties [9]. For example, Lee et al. [19] brazed Ti alloy and Al alloy using AlSi$_{10}$Mg filler. Chang et al. [20] brazed the Al6061 using Al-10.8Si-10Cu and Al-9.6Si-20Cu at 560 °C, and the results showed that the liquid phase line temperature changed from 592 °C to 570 °C when 10 wt. % Cu was added to the Al-12Si.

Contact reactive brazing is a kind of brazing method without any brazing flux [21], which has been widely applied for brazing Al alloys to other alloys, such as Al6063 [22], Al6061 to AZ31B Mg alloy [23], and Al6063 to 1Cr18Ni9Ti stainless steel [24]. Schällibaum et al. [25] studied the microstructure of the AA6082 brazed joints with plating copper, and the results showed that the formation of defects was caused by the residual oxide films aggregated in the brazed joint. Wu et al. [24] used Cu as an interlayer to join Al6063 and 1Cr18Ni9Ti stainless steel by contact reactive brazing. In fact, the existence of an oxide film on the surface of the aluminum alloy and titanium alloy prevented the diffusion and reaction during the brazing process, which deteriorated the interfacial microstructure and then reduced the joining properties [26]. Therefore, the appropriate surface treatment method should be adopted to remove the stable oxide film. As a method of surface modification, ion bombardment can effectively remove the oxide film [27,28], and our previous work also demonstrated it [26]. To prevent re-oxidation after the ion bombardment process, it was chosen that the Cu layers be prepared on their surfaces by magnetron sputtering for protection, as well as that the eutectic reaction between copper and aluminum would occur at 548 °C, which facilitated brazing of the contact reaction between TC4 and Al7075 at a relatively low temperature [29,30].

Based on our previous study, the combination of ion bombardment and magnetron sputtering copper deposition method was used to braze TC4 and Al7075 dissimilar alloys. The microstructural evolution of TC4/Al7075 brazed joints was discussed in detail under different brazing parameters (brazing temperature and holding time), and the TC4/Al7075 brazing processes were optimized based on the joining properties.

2. Experimental Procedures

The TC4 and Al7075 alloys were cut in a size of 15 mm × 10 mm × 5 mm and 8 mm × 8 mm × 5 mm, respectively. The brazing surfaces of the TC4 alloy and Al7075 alloy were ground with metallographic sandpaper and polished using a diamond agent down to 2.5 μm. Finally, the polished TC4 alloy and Al7075 alloy were cleaned with acetone under an ultrasonic bath and then air dried. The microstructure of the TC4 and Al7075 alloys are shown in Figure S1.

Figure 1a shows the schematic diagram of the entire process. The surface oxide film on the faying surfaces of the TC4 and Al7075 substrates was removed by Ar ion bombardment, and then a Cu layer with a thickness of 5 μm was deposited onto both sides of the brazing surface by magnetron sputtering [31,32]. Subsequently, the brazing process of Al7075 to TC4 was carried out in the furnace of a vacuum level of less than 5.0×10^{-3} Pa under the pressure of 0.25 MPa. For the brazing process, all assemblies were heated first to 535 °C at a heating rate of 10 °C/min, kept for 5 min, and then continually heated to the specified brazing temperature (560–620 °C) at a heating rate of 5 min/°C. Subsequently, the brazing samples were held for 15–60 min. Finally, the furnace was slowly cooled to room temperature.

The cross-sectional microstructure of the TC4/Al7075 joints was characterized by the field emission scanning electron microscopy (SEM, MERLIN Compact, Zeiss), energy dispersive spectrometer (EDS, OCTANE PLUS, EDAX), and X-ray diffraction (XRD, JDX-3530M). The shear strength of the TC4/Al7075 joints was tested at a constant rate of 0.5 mm/min by using a universal testing machine (Instron 5967) at room temperature (Figure 1b). The fracture mode and microstructure were analyzed using SEM equipped with EDS, and the phase of the fracture surfaces was identified by XRD.

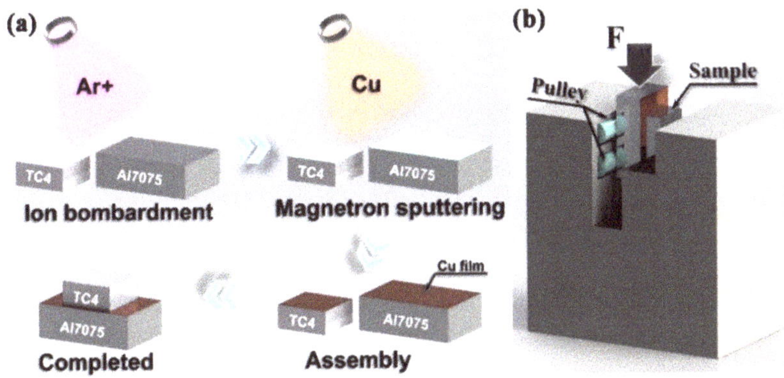

Figure 1. Schematic diagram of the TC4/Al7075 contact reactive brazing process (**a**) and shear test experiment (**b**).

3. Results and Discussion

3.1. Typical Microstructure of TC4/Cu Layer/Al7075 Brazed Joint

The typical microstructure of the TC4/Cu layer/Al7075 brazed joint at 600 °C for 30 min is shown in Figure 2. It can be seen that the Cu layer reacted completely with the base materials, and a sound joint was formed without any crack or void, as shown in Figure 2a. After the eutectic reaction between the Cu layer and the Al alloy, the resulting eutectic liquid phase penetrated the Al7075 substrate, and many intermetallic compounds (IMCs) were formed in the brazing seam. Spots A, B, C, D, E, and F represent the different phases of the brazed joint in Figure 2, and Table 1 shows the phase compositions of different spots determined by EDS. The atomic ratio of Al and Cu was 2:1 in spots A and E, which may be the Al_2Cu phase. The atomic ratio of Al and Ti was 3:1 in spot B, revealing the possible formation of the Al_3Ti phase based on the Al-Ti phase diagram (Figure S2) [33]. The atomic percent proportion of Al and Ti was approximately 5: 3 in spot C, which was confirmed as the Al_5Ti_3 phase [34]. Spot D with atomic percent proportion of Al, Cu, and Mg was approximately 2:1:1, which was speculated as the Al_2CuMg phase according to the Al-Cu-Mg ternary phase diagram (Figure S3) [35]. Spot F was analyzed as the Al-based solid solution (Al(s, s)) according to the EDS result.

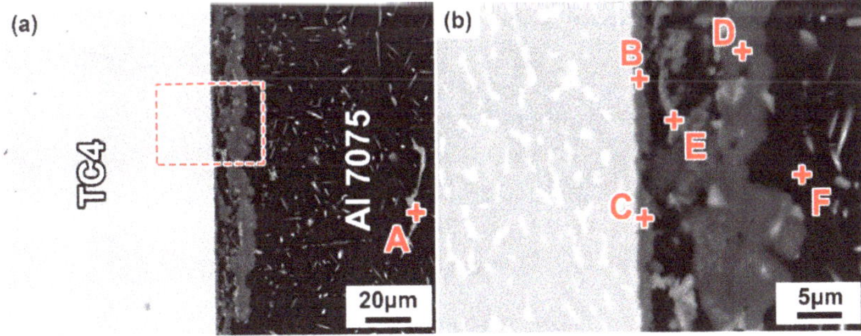

Figure 2. Interfacial microstructure of TC4/Cu layer/Al7075 alloy brazed joint at 600 °C for 30 min. (**a**) Low magnification; (**b**) High magnification

Table 1. EDS analysis of the selected spots in Figure 2 (at. %).

Spot	Al	Ti	Cu	Mg	Possible Phase
A	65.64	0.92	25.08	8.36	Al_2Cu
B	71.95	26.45	0.54	1.06	Al_3Ti
C	60.51	36.98	1.20	1.31	Al_5Ti_3
D	61.84	3.80	16.24	18.12	Al_2CuMg
E	63.60	2.30	28.57	5.53	Al_2Cu
F	96.47	1.22	1.17	1.14	$Al(s, s)$

The elemental distribution of the TC4/Al7075 joint is shown in Figure 3b–e. It can be seen that the substrate gradually dissolved into the eutectic liquid phase as the Al-Cu eutectic phase reacted and spread on the surface of Al alloy substrate during the brazing process. In addition, Figure 3b,e shows the concentrated distribution of Cu and Mg elements in the brazing seam. Okamoto et al. [36] demonstrated that Cu diffused into the Al substrate and formed Al_2Cu, which was also demonstrated from the Al-Cu binary phase diagram (Figure S4). Meanwhile, Figure 3e shows that the distribution of the Mg element primarily concentrated in the brazing seam and then formed the Al_2CuMg phase [15,32]. Liu et al. [37] confirmed the interface energy of Al_3Ti was the lowest, and preferentially formed on the Al substrate. In addition, the metastable intermediate phase of Al_5Ti_3 was formed during the reaction [38]. Therefore, the typical interfacial microstructure of the TC4/Al7075 joint brazed at 600 °C for 30 min was TC4 substrate/Al_3Ti + Al_5Ti_3/Al_2Cu + Al_2CuMg/Al7075 substrate.

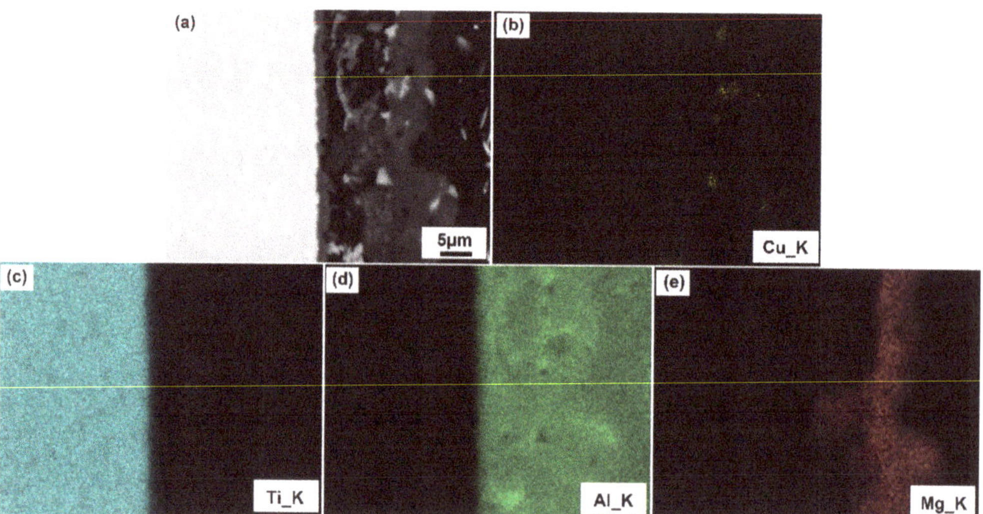

Figure 3. Elemental distribution of TC4/Cu layer/Al7075 brazed joint brazed at 600 °C for 30 min. (**a**) BSE image of the typical brazed joint and the elemental distribution of (**b**) Ti; (**c**) Al; (**d**) Cu; (**e**) Mg.

3.2. Effect of Brazing Parameters on the Microstructure of TC4/Cu Layer/Al7075 Brazed Joints

Figures 4 and 5 show the interfacial microstructures of the TC4/Cu layer/Al7075 brazed joints at various brazing parameters. All brazing temperatures were higher than the Al-Cu eutectic temperature (548 °C). When the brazing temperature was 560 °C (Figure 4a), large amounts of the Al_2Cu and Al_2CuMg phases formed in the brazing seam. With an increasing brazing temperature, the formed Al_2Cu and Al_2CuMg IMCs gradually decreased and disappeared due to the rapid diffusion of Cu atoms into Al7075 substrate. However, the grain coarsening of the Al7075 substrate appeared when the brazing temperature was

620 °C. When the temperature range was 560–620 °C, it is worth noting that the eutectic liquid phase mainly infiltrated along the Al grain boundaries, and the Al$_2$Cu phase formed at the grain boundaries of Al7075.

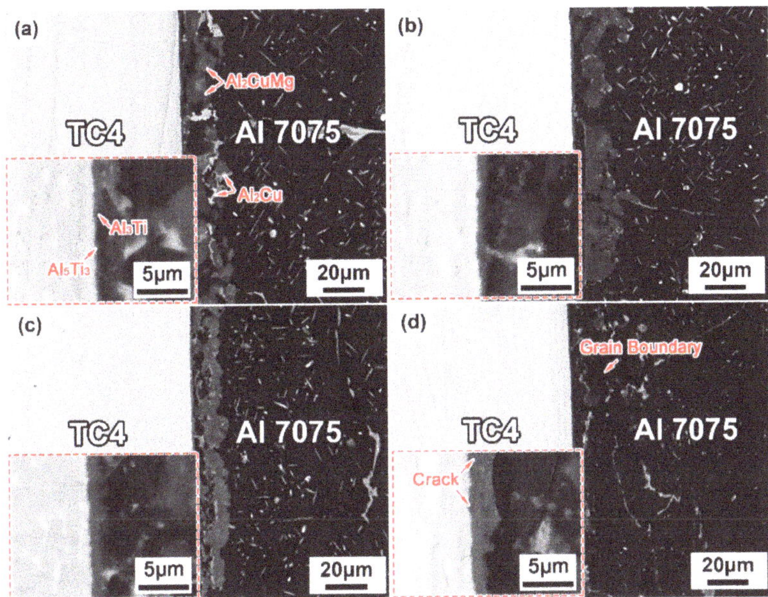

Figure 4. Interfacial microstructures of brazed joints at different temperatures for 30 min. (**a**) 560 °C; (**b**) 580 °C; (**c**) 600 °C; (**d**) 620 °C

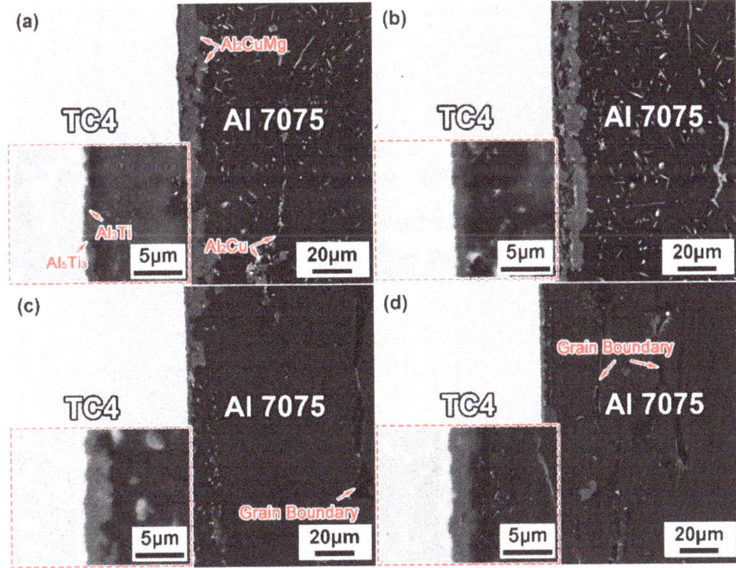

Figure 5. Interfacial microstructures of brazed joints at 600 °C for different holding times. (**a**) 15 min; (**b**) 30 min; (**c**) 45 min; (**d**) 60 min

Figure 5 illustrates the microstructural evolution of the brazed joints with prolonging the holding time. Insufficient diffusion of Cu led to the formation of large and continuous Al_2Cu and Al_2CuMg IMCs in the brazing seam at the short holding time (15 min). With the extension of the holding time (30 min), Cu atoms diffused fully into the Al substrates, which caused the decrease of IMCs. With the holding time further raised to 45 min or 60 min, the IMCs substantially reduced in the brazing seam, followed by the grain coarsening of Al7075, which worsened the properties of Al alloy.

In addition, as shown in Figures 4 and 5, a discontinuous reaction layer of thickness of less than 1 μm was formed on the TC4 side when the brazing parameters were insufficient. As the brazing parameters were raised, the intermetallic compounds' layer became thicker and more continuous. However, the microcracks appeared in the reaction layer when the brazing parameters were too high (brazing temperature ~ 620 °C and holding time 45–60 min), and it was presumed that it was caused by the difference in thermal expansion coefficient between the reaction layer and the base material. The corresponding EDS line scan results (Figure 6) indicated that the Al and Ti atoms diffused each other to form the diffusion layer.

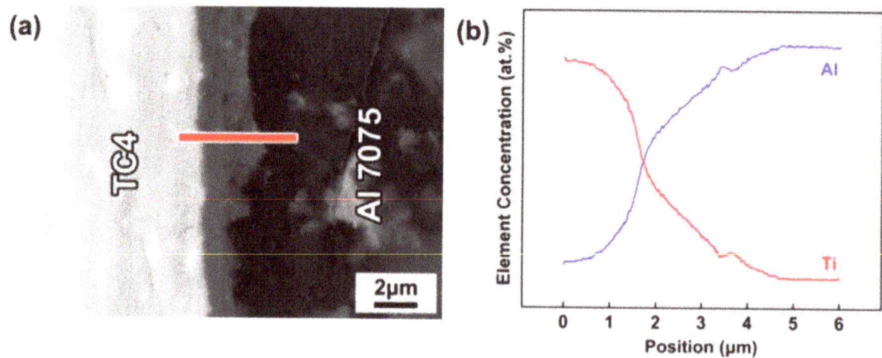

Figure 6. (a) Interfacial microstructure of brazed joint, **(b)** the EDS line scanning distribution of Ti and Al elements.

Based on the above analyses on the interfacial microstructure of the joints with different brazing parameters, the evolution of the TC4/Al7075 brazed joints can be proposed as follows. The Al-Cu eutectic liquid phase was formed when the brazing temperature exceeded the Al-Cu eutectic temperature of 548 °C. As the brazing parameters increased, the Cu atoms fully diffused into the substrate and reacted with Al to produce more eutectic liquid phase. Meanwhile, the Mg atoms from the substrate entered the liquid and reacted with Al and Cu atoms to form Al_2CuMg. The residual liquid solidified and formed the eutectic structure (α-Al + Al_2Cu) in the brazing seam during the cooling stage [32]. On the other hand, the formation of the Al-Cu eutectic liquid phase could promote the diffusion of Ti atoms into the liquid phase and produce the diffusion gradient on the TC4 side. The Al and Ti elements reacted and formed Al_3Ti according to the Al-Ti binary phase diagram [15]. Moreover, the metastable intermediate phase of Al_5Ti_3 was formed in the interface during the reaction, owing to insufficient atomic diffusion [38].

3.3. Effect of Brazing Parameters on the Mechanical Properties of Brazed Joints

Figure 7 shows the shear strength of the TC4/Cu layer/Al7075 brazed joints at various brazing parameters. The shear strength improved first and then decreased evidently, and the maximum shear strength of ~201.45 ± 4.4 MPa was obtained at 600 °C for 30 min. Combined with the analysis of the interfacial microstructures, large amounts of brittle intermetallic compounds of the Al_2Cu and Al_2CuMg phase distributed continuously in the brazing seam when the brazing temperature was low (560 °C), which deteriorated the joining properties. The enhanced diffusion ability of the Cu layer led to the reduction of Al-Cu IMCs, and the composition of the brazing seam tended to be uniform at a brazing temperature of 600 °C and a holding time of 30 min, which enhanced the joint strength effectively. However, with a further increase of the brazing temperature (620 °C) and holding time (45 min and 60 min), the sufficient diffusion of Cu atoms resulted in the decrease of the Al-Cu intermetallic compound, and the grain growth of the Al substrate and the formed microcracks at grain boundaries had a detrimental effect on the shear strength.

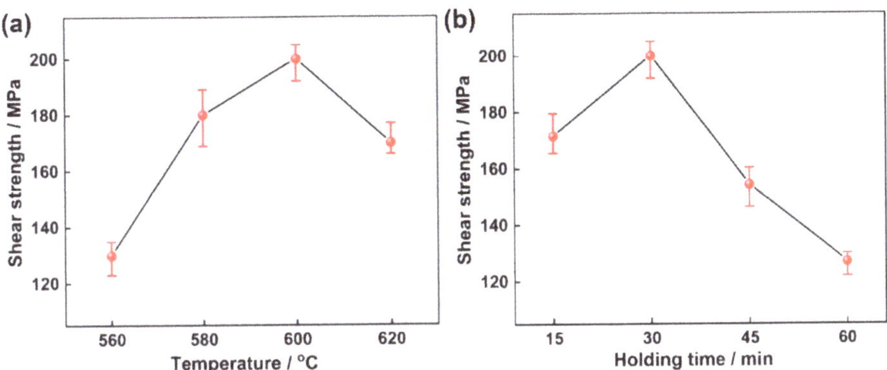

Figure 7. Effect of the brazing parameters on mechanical properties of brazed joints. (**a**) Brazing temperature; (**b**) Holding time.

To further analyze the effect of the brazing parameters on the fracture path and fracture mode of the joints, the fracture analysis of the joints was performed after the properties' test, as shown in Figures 8 and 9. As the brazing parameters increased, the cracks extended mainly along the intermetallic compounds. The XRD pattern of the fracture surface (Figure 10) showed the presence of Al_2Cu, Al_2CuMg, Al_5Ti_3, and Al_3Ti phases, which was consistent with the above results. With a further increase of the brazing parameters, the Al-Cu intermetallic compounds decreased gradually, and the fracture path mainly propagated along the intermetallic compounds in the brazing seam. Except for that, part of the cracks propagated inside the Al grains and a transgranular fracture formed. With the continuous elevation of the brazing parameters, the contents of Al_2Cu and Al_2CuMg intermetallic compounds decreased gradually. Meanwhile, the cracks propagated inside the Al substrates and IMCs. When the brazing parameters were too high (above 600 °C), the fracture extended along the Al grain and intergranular fracture happened.

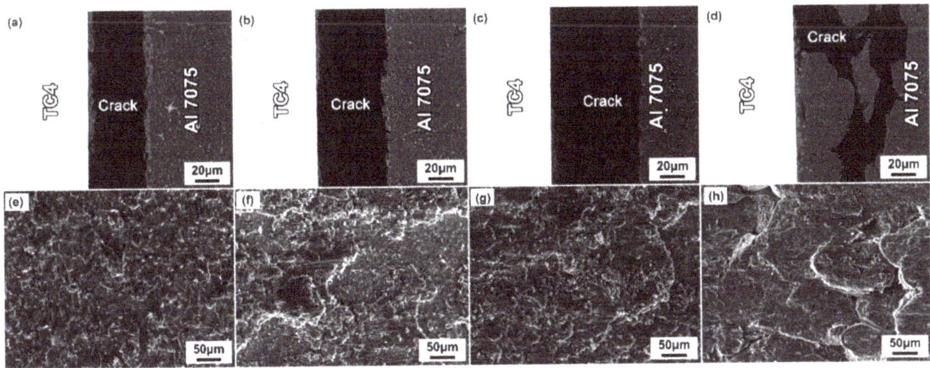

Figure 8. Fracture morphologies of TC4/Cu layer/Al7075 joints brazed at different temperatures for 30 min. (**a,e**) 560 °C; (**b,f**) 580 °C; (**c,g**) 600 °C; (**d,h**) 620 °C.

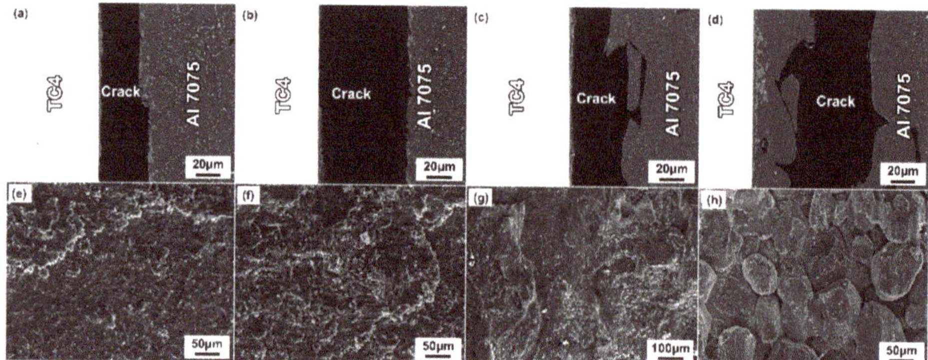

Figure 9. Fracture morphologies of TC4/Cu layer/Al7075 joints brazed at 600 °C for different times. (**a,e**) 15 min; (**b,f**) 30 min; (**c,g**) 45 min; (**d,h**) 60 min.

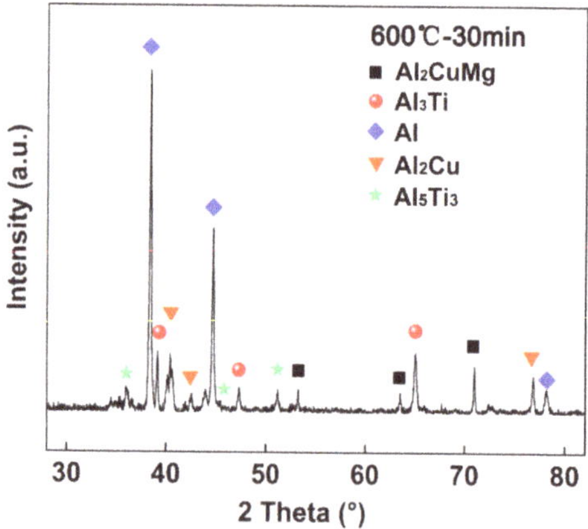

Figure 10. XRD pattern of the fracture surface (Al7075 side).

4. Conclusions

1. The contact reactive brazing of the TC4 alloy to the Al7075 alloy was achieved using deposited Cu as an interlayer. The typical interfacial microstructure of the TC4/Al7075 brazed joint was the TC4 substrate/Al$_3$Ti + Al$_5$Ti$_3$/Al$_2$Cu + Al$_2$CuMg/Al7075 substrate at 600 °C for 30 min.
2. With increasing the brazing temperature and holding time, the amount of Al$_2$Cu and Al$_2$CuMg IMCs in the brazed joints decreased and the homogenization of the joint composition improved, while the thickness of the reaction layer (Al$_3$Ti + Al$_5$Ti$_3$) on the TC4 side increased gradually.
3. The shear strength improved first and then decreased with increasing brazing parameters, and the maximum shear strength of ~201.45 ± 4.40 MPa was obtained at 600 °C for 30 min. The fracture mode of the joint changed from brittle fracture to transgranular fracture, and the intergranular fracture occurred when the brazing temperature was higher than 600 °C and the holding time exceeded 30 min.

Supplementary Materials: The following are available online at https://www.mdpi.com/article/10.3390/ma14216570/s1, Figure S1: the characterization microstructures of the TC4 and Al7075 substrate: (a) TC4; (b) Al7075, Figure S2: Al-Ti binary phase diagram, Figure S3: Al-Cu-Mg ternary phase diagram, Figure S4: Al-Cu binary phase diagram.

Author Contributions: Writing—original draft preparation, M.Y.; Conceptualization, M.Y. and C.N.; Methodology, S.H.; Validation, M.Y., C.N. and S.H.; Formal analysis, X.S.; Investigation, M.Y. and C.N.; Resources, Y.P. and W.L.; Writing—review and editing, S.H.; Supervision, S.H. and J.Z.; Funding acquisition, X.S. All authors have read and agreed to the published version of the manuscript.

Funding: This work was funded from the National Natural Science Foundation of China (Grant Nos. 51905125 and 51775138), the Taishan Scholars Foundation of Shandong Province (No. tsqn 201812128), and the Natural Science Foundation of Shandong Province (No. ZR2019BEE031).

Institutional Review Board Statement: Not applicable.

Informed Consent Statement: Not applicable.

Data Availability Statement: The raw/processed data in the paper cannot be shared at present owing to part of an ongoing further study.

Conflicts of Interest: The authors declare no conflict of interest.

References

1. Emadinia, O.; Guedes, A.; Tavares, C.; Simões, S. Joining Alumina to Titanium Alloys Using Ag-Cu Sputter-Coated Ti Brazing Filler. *Materials* **2020**, *13*, 4802. [CrossRef]
2. Yang, X.; Wang, Y.; Dong, X.; Peng, C.; Ji, B.; Xu, Y.; Li, W. Hot deformation behavior and microstructure evolution of the laser solid formed TC4 titanium alloy. *Chin. J. Aeronaut.* **2020**, *34*, 163–182. [CrossRef]
3. Dong, H.; Yang, Z.; Wang, Z.; Deng, D.; Dong, C. Vacuum Brazing TC4 Titanium Alloy to 304 Stainless Steel with Cu-Ti-Ni-Zr-V Amorphous Alloy Foil. *J. Mater. Eng. Perform.* **2014**, *23*, 3770–3777. [CrossRef]
4. AlHazaa, A.; Alhoweml, I.; Shar, M.A.; Hezam, M.; Abdo, H.S.; AlBrithen, H. Transient Liquid Phase Bonding of Ti-6Al-4V and Mg-AZ31 Alloys Using Zn Coatings. *Materials* **2019**, *12*, 769. [CrossRef]
5. Lan, J.; Chen, Z.; Liu, L.; Zhang, Q.; He, M.; Li, J.; Peng, X.; Fan, T. The Thermal Properties of L1$_2$ Phases in Aluminum Enhanced by Alloying Elements. *Metals* **2021**, *11*, 1420. [CrossRef]
6. Alhazaa, A.; Khan, T.; Haq, I. Transient liquid phase (TLP) bonding of Al7075 to Ti–6Al–4V alloy. *Mater. Charact.* **2010**, *61*, 312–317. [CrossRef]
7. Dong, Y.; Zhang, C.; Luo, W.; Yang, S.; Zhao, G. Material flow analysis and extrusion die modifications for an irregular and multitooth aluminum alloy radiator. *Int. J. Adv. Manuf. Technol.* **2016**, *85*, 1927–1935. [CrossRef]
8. Chang, S.; Tsao, L.; Lei, Y.; Mao, S.; Huang, C. Brazing of 6061 aluminum alloy/Ti–6Al–4V using Al–Si–Cu–Ge filler metals. *J. Mater. Process. Technol.* **2012**, *212*, 8–14. [CrossRef]
9. Wang, Z.; Shen, J.; Hu, S.; Wang, T.; Bu, X. Investigation of welding crack in laser welding-brazing welded TC4/6061 and TC4/2024 dissimilar butt joints. *J. Manuf. Process.* **2020**, *60*, 54–60. [CrossRef]
10. Gao, M.; Chen, C.; Gu, Y.; Zeng, X. Microstructure and Tensile Behavior of Laser Arc Hybrid Welded Dissimilar Al and Ti Alloys. *Materials* **2014**, *7*, 1590–1602. [CrossRef]
11. Baqer, Y.M.; Ramesh, S.; Yusof, F.; Manladan, S.M. Challenges and advances in laser welding of dissimilar light alloys: Al/Mg, Al/Ti, and Mg/Ti alloys. *Int. J. Adv. Manuf. Technol.* **2018**, *95*, 4353–4369. [CrossRef]
12. Schubert, E.; Klassen, M.; Zerner, I.; Walz, C.; Sepold, G. Light-weight structures produced by laser beam joining for future applications in automobile and aerospace industry. *J. Mater. Process. Technol.* **2001**, *115*, 2–8. [CrossRef]
13. Casalino, G.; D'Ostuni, S.; Guglielmi, P.; Leo, P.; Mortello, M.; Palumbo, G.; Piccininni, A. Mechanical and microstructure analysis of AA6061 and Ti6Al4V fiber laser butt weld. *Optik* **2017**, *148*, 151–156. [CrossRef]
14. Song, Z.; Nakata, K.; Wu, A.; Liao, J. Interfacial microstructure and mechanical property of Ti6Al4V/A6061 dissimilar joint by direct laser brazing without filler metal and groove. *Mater. Sci. Eng. A* **2012**, *560*, 111–120. [CrossRef]
15. Naeimian, H.; Mofid, M.A. TLP bonding of Ti–6Al–4V to Al 2024 using thermal spray Babbitt alloy interlayer. *Trans. Nonferrous Met. Soc. China* **2020**, *30*, 1267–1276. [CrossRef]
16. Takemoto, T.; Okamoto, I. Intermetallic compounds formed during brazing of titanium with aluminium filler metals. *J. Mater. Sci.* **1988**, *23*, 1301–1308. [CrossRef]
17. Assari, A.H.; Eghbali, B. Solid state diffusion bonding characteristics at the interfaces of Ti and Al layers. *J. Alloys Compd.* **2018**, *773*, 50–58. [CrossRef]
18. Chandrappa, K.; Kant, R.; Ali, R.; Vineth, K. Optimization of process parameter of diffusion bonding of Ti-Al and Ti-Cu. *Mater. Today Proc.* **2020**, *27*, 1689–1695. [CrossRef]
19. Lee, T.W.; Kim, I.K.; Lee, C.H.; Kim, J.H. Growth behavior of intermetallic compound layer in sandwich-type Ti/Al diffusion couples inserted with Al-Si-Mg alloy foil. *J. Mater. Sci. Lett.* **1999**, *18*, 1599–1602. [CrossRef]
20. Chang, S.; Tsao, L.; Li, T.; Chuang, T. Joining 6061 aluminum alloy with Al–Si–Cu filler metals. *J. Alloys Compd.* **2009**, *488*, 174–180. [CrossRef]
21. Tan, F.F.; Du, K. Analysis of Organizations of Brazed Seam 5052 Al Alloy Contact Reactive Brazing. *Adv. Mater. Res.* **2013**, *690-693*, 2598–2600. [CrossRef]

22. Wu, M.F.; Yu, C.; Pu, J. Study on microstructures and grain boundary penetration behaviours in contact reactive brazing joints of 6063 Al alloy. *Mater. Sci. Technol.* **2008**, *24*, 1422–1426. [CrossRef]
23. Liu, L.; Tan, J.; Liu, X. Reactive brazing of Al alloy to Mg alloy using zinc-based brazing alloy. *Mater. Lett.* **2007**, *61*, 2373–2377. [CrossRef]
24. Wu, M.-F.; Si, N.-C.; Chen, J. Contact reactive brazing of Al alloy/Cu/stainless steel joints and dissolution behaviors of interlayer. *Trans. Nonferrous Met. Soc. China* **2011**, *21*, 1035–1039. [CrossRef]
25. Schällibaum, J.; Burbach, T.; Münch, C.; Weiler, W.; Wahlen, A. Transient liquid phase bonding of AA 6082 aluminium alloy. *Mater. Werkst.* **2015**, *46*, 704–712. [CrossRef]
26. Niu, C.; Han, J.; Hu, S.; Song, X.; Long, W.; Liu, D.; Wang, G. Surface modification and structure evolution of aluminum under argon ion bombardment. *Appl. Surf. Sci.* **2020**, *536*, 147819. [CrossRef]
27. Park, M.; Baek, S.; Kim, S.; Kim, S.E. Argon plasma treatment on Cu surface for Cu bonding in 3D integration and their characteristics. *Appl. Surf. Sci.* **2015**, *324*, 168–173. [CrossRef]
28. Kim, T.H.; Howlader, M.M.R.; Itoh, T.; Suga, T. Room temperature Cu–Cu direct bonding using surface activated bonding method. *J. Vac. Sci. Technol. A* **2003**, *21*, 449–453. [CrossRef]
29. Niu, C.; Song, X.; Hu, S.; Lu, G.; Chen, Z.; Wang, G. Effects of brazing temperature and post weld heat treatment on 7075 alloy brazed joints. *J. Mater. Process. Technol.* **2018**, *266*, 363–372. [CrossRef]
30. Song, X.; Niu, C.; Hu, S.; Liu, D.; Cao, J.; Feng, J. Contact reactive brazing of Al7075 alloy using Cu layer deposited by magnetron sputtering. *J. Mater. Process. Technol.* **2018**, *252*, 469–476. [CrossRef]
31. Niu, C.; Han, J.; Hu, S.; Chao, D.; Song, X.; Howlader, M.; Cao, J. Fast and environmentally friendly fabrication of superhydrophilic-superhydrophobic patterned aluminum surfaces. *Surfaces Interfaces* **2020**, *22*, 100830. [CrossRef]
32. Hu, S.; Niu, C.; Bian, H.; Song, X.; Cao, J.; Tang, D. Surface-activation assisted brazing of Al-Zn-Mg-Cu alloy: Improvement in microstructure and mechanical properties. *Mater. Lett.* **2018**, *218*, 86–89. [CrossRef]
33. Pripanapong, P.; Kariya, S.; Luangvaranunt, T.; Umeda, J.; Tsutsumi, S.; Takahashi, M.; Kondoh, K. Corrosion Behavior and Strength of Dissimilar Bonding Material between Ti and Mg Alloys Fabricated by Spark Plasma Sintering. *Materials* **2016**, *9*, 665. [CrossRef] [PubMed]
34. Hayashi, K.; Nakano, T.; Umakoshi, Y. Meta-stable region of Al5Ti3 single-phase in time-temperature-transformation (TTT) diagram of Ti–62.5 at.% Al single crystal. *Intermetallics* **2002**, *10*, 771–781. [CrossRef]
35. Chen, S.-L.; Zuo, Y.; Liang, H.; Chang, Y.A. A thermodynamic description for the ternary Al-Mg-Cu system. *Met. Mater. Trans. A* **1997**, *28*, 435–446. [CrossRef]
36. Okamoto, H. *Desk Handbook: Phase Diagrams for Binary Alloys*; ASM International: Geauga, OH, USA, 2010; p. 315.
37. Liu, J.; Su, Y.; Xu, Y.; Luo, L.; Guo, J.; Fu, H. First Phase Selection in Solid Ti/Al Diffusion Couple. *Rare Met. Mater. Eng.* **2011**, *40*, 753–756. [CrossRef]
38. Huang, J.; Liu, Y.; Liu, S.; Guan, Z.; Yu, X.; Wu, H.; Yu, S.; Fan, D. Process of welding-brazing and interface analysis of lap joint Ti-6Al-4V and aluminum by plasma arc welding. *J. Manuf. Process.* **2020**, *61*, 396–407. [CrossRef]

Article

Prediction Model of the Resulting Dimensions of Welded Stamped Parts

Milan Kadnár [1], Peter Káčer [1], Marta Harničárová [2,3,*], Jan Valíček [2,3], Mirek Gombár [4], Milena Kušnerová [3], František Tóth [1], Marian Boržan [5] and Juraj Rusnák [1]

[1] Department of Machine Design, Faculty of Engineering, Slovak University of Agriculture in Nitra, Tr. A. Hlinku 2, 949 76 Nitra, Slovakia; milan.kadnar@uniag.sk (M.K.); kacer.pet@gmail.com (P.K.); frantisek.toth@uniag.sk (F.T.); juraj.rusnak@uniag.sk (J.R.)

[2] Department of Electrical Engineering, Automation and Informatics, Faculty of Engineering, Slovak University of Agriculture in Nitra, Tr. A. Hlinku 2, 949 76 Nitra, Slovakia; jan.valicek@uniag.sk

[3] Department of Mechanical Engineering, Faculty of Technology, Institute of Technology and Business in České Budějovice, Okružní 10, 370 01 České Budějovice, Czech Republic; kusnerova.milena@mail.vstecb.cz

[4] Faculty of Management, University of Prešov in Prešov, Konštantínova 16, 080 01 Prešov, Slovakia; miroslav.gombar@unipo.sk

[5] Department of Manufacturing Engineering, Faculty of Machine Building, Technical University of Cluj-Napoca, B-dul Muncii, No. 103-105, 400641 Cluj-Napoca, Romania; mborzan@yahoo.com

* Correspondence: marta.harnicarova@uniag.sk; Tel.: +421-37-641-5782

Abstract: The combination of stamping and subsequent welding of components is an important area of the automotive industry. Stamping inaccuracies affect the final size of the stamping and the welded part. In this article, we deal with a specific component that is produced by such a procedure and is also a common part of the geometry of a car. We focused on the possibility of using a negative phenomenon—deformation during welding—on the partial elimination of inaccuracies arising during stamping. Based on the planned experiment, we created a prediction model for the selected part and its production, with the help of which it is possible to determine suitable welding parameters for a specific dimension of the stamping and the required monitored dimension of the welded part. The article also includes the results of additional experimental measurements verifying the accuracy of the model and prediction maps for practice.

Keywords: welding; distortion; stamping; model; prediction

Citation: Kadnár, M.; Káčer, P.; Harničárová, M.; Valíček, J.; Gombár, M.; Kušnerová, M.; Tóth, F.; Boržan, M.; Rusnák, J. Prediction Model of the Resulting Dimensions of Welded Stamped Parts. *Materials* **2021**, *14*, 3062. https://doi.org/10.3390/ma14113062

Academic Editors: Drazan Kozak, Nenad Gubeljak and Aleksandar Sedmak

Received: 3 May 2021
Accepted: 2 June 2021
Published: 3 June 2021

Publisher's Note: MDPI stays neutral with regard to jurisdictional claims in published maps and institutional affiliations.

1. Introduction

Welding is the most important method of joining metal components. There are many joining techniques available. Each technology is specific and has its advantages or disadvantages for a certain application. The sectors where welding plays a dominant role include the automotive industry. The most common welding techniques in the automotive industry are laser beam welding [1], metal inert gas welding process, metal active gas welding process [2], and spot welding [3].

MIG (metal inert gas) welding process was used in this study. MIG welding is a remarkably flexible method. The principle of MIG welding is that the melted bath is protected from the effects of the surrounding atmosphere (mainly oxygen and nitrogen) by a protective atmosphere, which may be inert or active. Inert atmospheres do not enter into chemical reactions with the melting bath. Active atmospheres participate in chemical reactions in the melting bath; their action is being compensated by a suitable composition of the additive material [4,5]. Weld heat input is important, because it affects the amount of distortion and residual stress in the component. The problems of these thermal effects significantly affect manufacturers in the automotive industry, because the welded assembly of the car must be kept in tight tolerances [6].

The vehicle body is created by several stamped components, which are joined together by the spot-welding process. The quality of the joint is determined by its geometry as well as by its local properties (mechanical, microstructural, chemical, etc.). Problems with general weldability can be solved with the right choice of materials suitable for welding, the design of welding technology suitable for the selected material, and the implementation of appropriate design modifications needed for successful welding. It follows that weldability is influenced by three main groups of interrelated factors, material–design–technology. The interconnection of the three main factors cannot be divided and assessed independently but must always be assessed comprehensively. In manufacturing a car, a combination of different technological processes often takes place, whereas a considerable number of factors influence the resulting parameters of the manufactured component. The typical combination of pressing and subsequent welding of stamped parts is an area that still deserves increased attention. Cost optimisation and efficient use of technological equipment in the case of small and medium-sized enterprises lead to an operational change in production. In the case of pressing, this means that a tool is often changed to ensure the production of different products. When changing tools, height adjustment deviations can occur due to imperfectly clean loading surfaces and mounting clearances. These directly affect the height of the press and, thus, the size of the stamped part [7]. Steel coils, where the changes in the chemical composition occur, also have a significant effect on the dimensions. Therefore, in the subsequent welding operation, the size of the stamped part represents the input factor [8]. The major causes of the dimensional inconsistencies in vehicle body components can be classified in the following points: assembly operations and positional capabilities, material properties (including the history of material production), stamping process parameters, and the welding process, itself (Figure 1) [9].

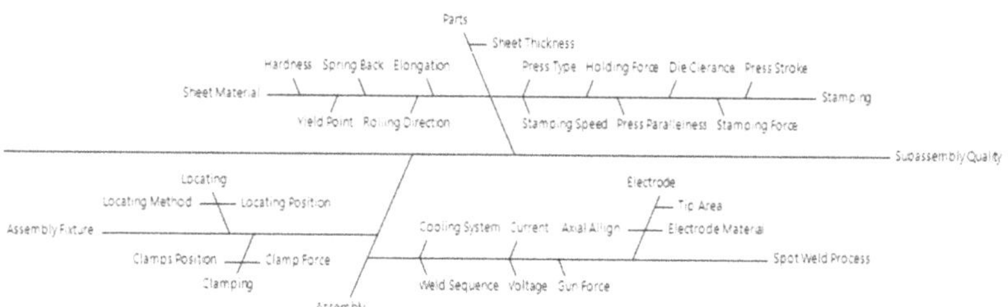

Figure 1. Ishikawa diagram for the identification of the subassembly quality of a car [9].

Over the last several years, there has been a rising demand for quality in the automotive industry, which forces manufacturers to solve considerable problems [10]. One of them is the assembly and setting of wheel geometry parameters. When the prescribed dimensional values are not observed, it can lead to problems that are reflected in the difficult or impossible deflection/tilt setting. Negative/positive tilt/deflection negatively affect the driving characteristics of the car. The car is not suitable for sale, but it is intended for repair. The dimensional deviations of certain parts and their joints affect the geometry the most. In this paper, we deal with the use of the negative phenomenon of deformation during welding to eliminate the differences caused by pressing a particular component, which is part of the geometry of a car (Figure 2).

Several studies have been made to study the effects of welding process parameters on the resulting geometry. Hamedi et al. [9] studied the effects of spot-welding parameters (current, time, and gun force) on the deformation of the subassemblies and the overall quality of the car body. They used a neural network and multi-objective genetic algorithm to find out the optimum values in order to get the least values of dimensional deviations in the subassemblies. Similar work was done by Kim et al. [11], who dealt with the response surface methodology to optimise the welding current, welding time, and welding force as input parameters and shear strength and indentation as output parameters. The finite element method (FEM) has been used for this purpose by Caro et al. [12]. The sheet metal forming procedure of a double-curved component made of alloy 718 has been studied using the FE method. This approach seems to be suitable for predicting distortions located in the same places as found during the experiment. Thomas et al. [7] confirmed the presumption of suitability for the use of the finite element method for accurate predictions of the final shape of stamped automotive assemblies, including the springback deformation of parts. Li et al. [13] studied the relationship between the weld quality and various process conditions using a two-stage, sliding-level experiment. A detailed description of the statistical analysis is shown in Zhang et al. [14] for predictions of expulsion limits. Muthu [4] performed an analysis of variance (ANOVA) to find the most significant parameters affecting the spot weld quality characteristics. An alternative approach based on the Taguchi method was used to analyse every welding process parameter for obtaining optimal weld pool geometry [15]. Other researchers [16,17] have attempted to find the optimal welding parameters by artificial neural networks (ANN).

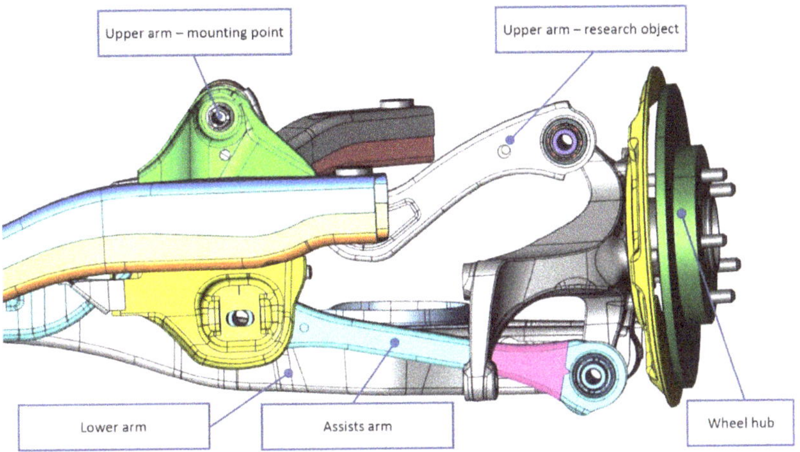

Figure 2. Alignment corrections.

Welding process failures and dimensional changes are the main leading factors to a decrease in productivity. Deformation induced by welding has many negative effects, one of which affects the dimension accuracy. The aim is to minimise the welding deformation. To mitigate this problem, the predictions play an important role to improve manufacturing accuracy. Statistical process control is often used to control the process in order to keep a high dimensional quality of the product [18].

Most published papers are focused on using the finite element method (FEM) to perform engineering analysis. FEA models have an advantage: they consider the effects during the welding, such as the phase transformations and the transformation strains during cooling. Some procedures of a dimensional control in the full automotive body are shown in [19–21]. The progress has been made also by developing the structure analysis method [22], the knowledge-based and model-based diagnostics techniques or tolerance analysis based on a mechanistic model [23].

Based on the literature search, it can be stated that the use of deformation during welding to eliminate inaccuracies caused by pressing has not been addressed so far. Therefore, in the paper, we present the original results of welded part deformation measurements (Donghee Slovakia, Ltd., Strečno, Slovakia), whose complex shape and changing dimensions limit the possibility of determining the optimal welding parameters. The influence of the basic process parameters, namely the welding current and the welding speed in combination with the changing size of the stamping, was verified. The presented original methodology and prediction model in practice allow welders to control the final dimension of the welded part with great accuracy and to respond to dimensional or capacity requirements of the production.

2. Materials and Methods

The object of research is a welded part (2 stampings)—see an example in Figure 3. Its thickness is 24 mm, and the dimension to be checked is supposed to be 315.5 ± 0.2 mm. In this particular case, the whole series of stampings is 0.68 mm larger than the nominal value.

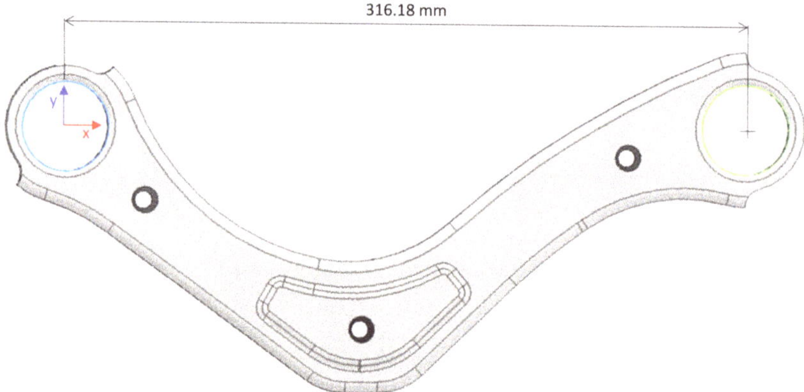

Figure 3. A welded part and the dimension to be examined.

The stampings are stamped on a SIMPAC MC2-500 press using a progressive 13-operation mould (Figure 4). The size of the stamping is affected by the setting of the press and the tool, and, during the series, the size is constant and regularly checked. After switching to a new series, the size of the stamping changes directly affects the resulting size of the welded part.

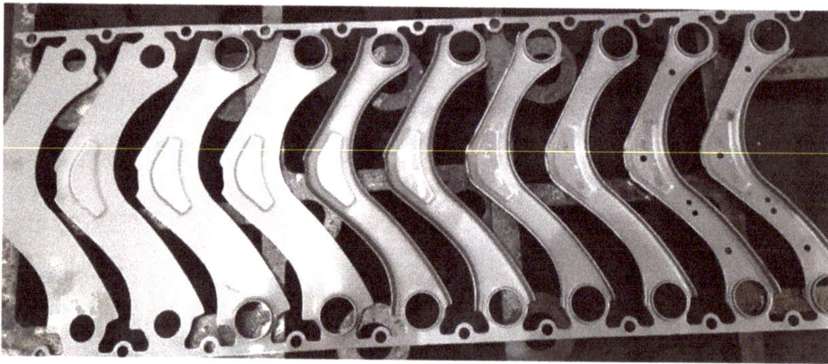

Figure 4. Stamping procedure.

All examined samples were stamped from Dual phase-type steel SGAFC590DP. Chemical properties are given in Table 1 and mechanical properties in Table 2. The values were obtained from the material sheet from the company Hyundai Steel Co., Ltd. (Seoul, Korea).

As with any process, the welding process is significantly affected by the process setting parameters. There are a number of associated parameters with this technology, among which the current, voltage, and welding speed are significant and precisely controllable. The OTC DM-400 welding machine (OTC Daihen Europe, GmbH., Mönchengladbach, Germany) deployed in combination with the ALmega AX V6 welding robot (OTC Daihen Europe, GmbH., Mönchengladbach, Germany) offers the possibility of automatic voltage determination. Since this is used in practice, we decided to consider this parameter constant (due to the multi-collinearity of the model). In the experiment, we considered both current and welding speed to be the variables. A summary of welding parameters is given in Table 3.

Table 1. Chemical composition (weight %) of the tested material.

Tested Material	C (%)	Si (%)	Mn (%)	P (%)	S (%)
SGAFC590DP (2 mm thickness)	0.071	0.183	1.895	0.018	0.004

Table 2. Mechanical properties of the tested material.

Tested Material	Yield Strength (MPa)	Ultimate Tensile Strength (MPa)	Elongation (%)
SGAFC590DP (2 mm thickness)	405	643	28

Table 3. Welding parameters.

Parameter	Parameter Type	Marking	Unit
Current	Variable	I	A
Welding speed		v	$cm \cdot min^{-1}$
Voltage	Automatic	U	V
Gas/wire dosing		–	–
Technology	Constant	MIG	–
Shielding welding gas		Ar	–
Wire-diameter		$d = 1.2$	mm
Wire-type		KISWEL KC-25M	–
Location and order of welds		–	–
Clamping parts		–	–

The location of the welds and their order during welding are shown in Figure 5 and the geometrical shape of lap weld joint in Figure 6.

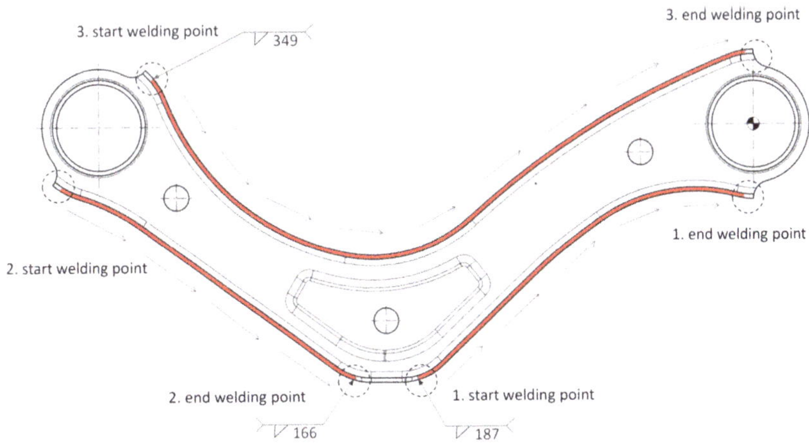

Figure 5. Location of welds and their order during welding.

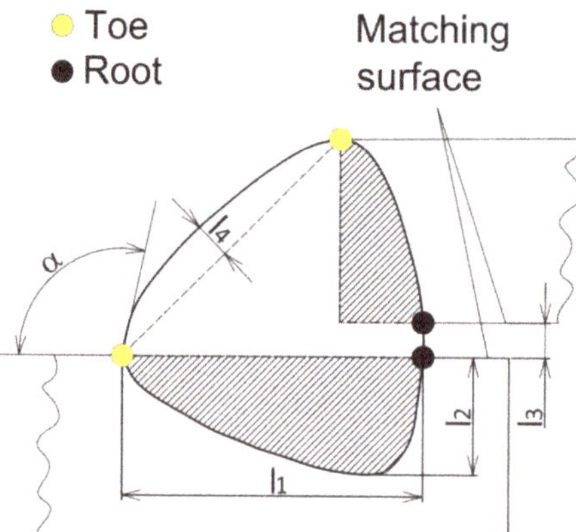

Figure 6. Geometrical shape of weld joint with the monitored dimensions in quality inspection (α—toe angle; l_1—leg length; l_2—penetration; l_3—gap width; l_4—excess welding).

The parameters used during the measurements were given by a combination of welding speed from 50–70 cm·min^{-1} and electric current 160–200 A, while each change in current also resulted in a change in the corresponding voltage (automatically). The currents 180–200 A are the most used in this production setting due to the lower error rate and, at the same time, due to the possible higher welding speeds.

Lower current values, i.e., 160–180 A proved to be equally applicable; the values lower than 150 A required a lower welding speed due to the correct weld of the material, which is not applicable in technical practice. From the point of view of the experiment, it was also important that the deformations were minimal to zero at the level lower than 150 A in the preliminary tests. Sections of welds realized under the boundary conditions of the experiment from the point of view of the introduced heat are shown in Figure 7.

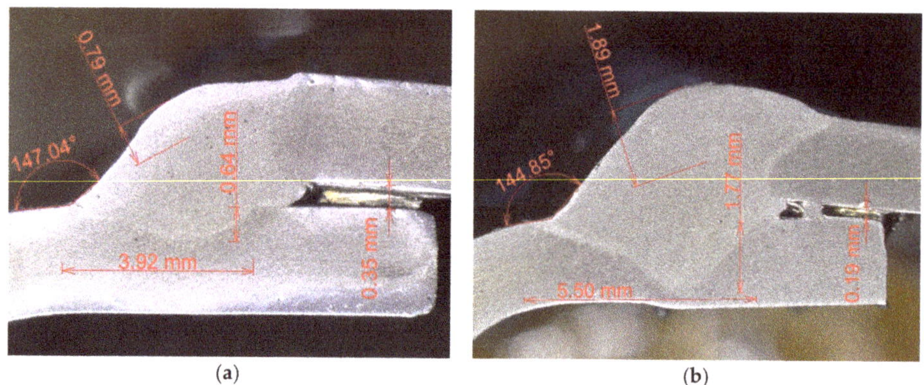

Figure 7. Weld cut: (**a**) Welding parameters I = 160 A, v = 70 cm·min^{-1}; (**b**) welding parameters I = 200A, v = 50 cm·min^{-1}.

The total deviation of the component (against 3D model, measured on a Romer Absolute Arm device) before and after welding are shown in Figures 8 and 9. The colour map represents the spatial deformation, and the dimension is monitored.

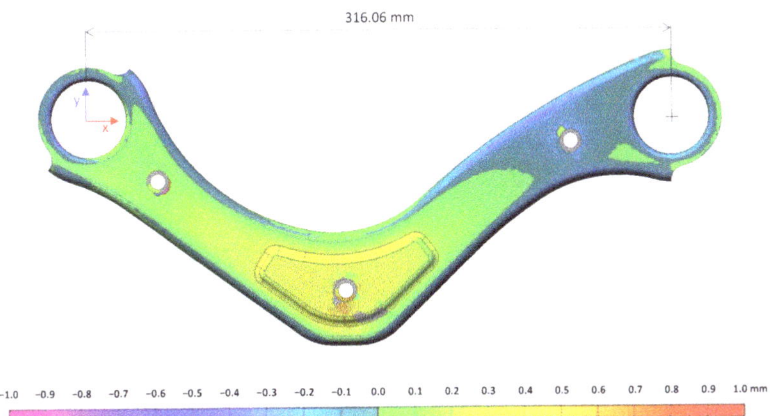

Figure 8. The total deviation of the stamped component from the 3D model.

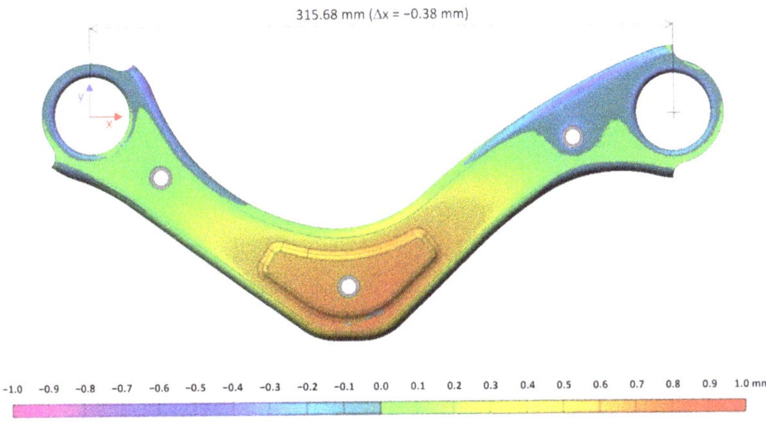

Figure 9. The total deviation of the welded part from the 3D model. Welding parameters (I = 190 A, v = 60 cm·min^{-1}).

Currents of a bigger magnitude than 200 A cause a high error rate (burn-trough) since, at high currents, a high speed must be selected, which also places great demands on accuracy. This cannot be achieved with the current technologies used. When tested at these levels, the results were very unstable, and the welds were of poor quality (holes in the welds, Figure 10), which had a negative effect on the overheating of the material and also on the deformation.

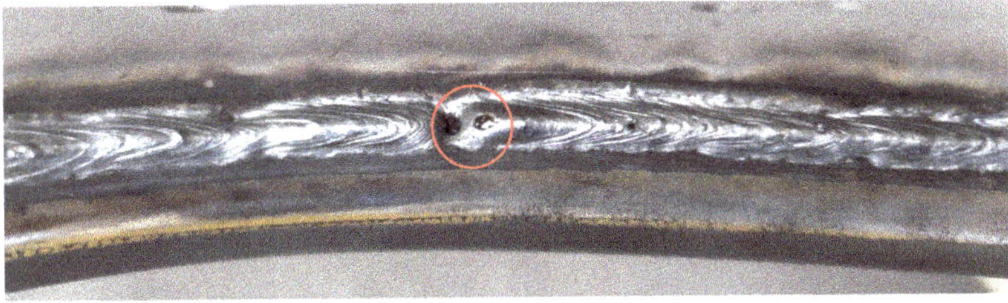

Figure 10. Defects when using incorrect welding parameters (I = 210 A, v = 80 cm·min^{-1}).

In order to determine the influence of selected factors on the final dimension of the welded part, an experiment was performed, and statistical analysis of the experimentally obtained data was performed. The influence of three technological factors such as welding current, welding speed, and stamping size were examined. The experiment was carried out according to a partial central composite design, and, due to the significant effect of axial points on the resulting weld quality, the Face Centered variant of axial points was chosen. Pseudo-central points were also part of the plan, as it is not possible to provide measurements at the central level for factor x_3 for operational reasons. Levels of factor x_3 were selected based on long-term observations at the lower and upper limit of the produced stampings. The levels of factors x_1 and x_2 were chosen with respect to the penetration tests of welds performed during long-term observations, the standard values of these factors, and the desired effect—reducing the size of the welded part. The selected factor levels are given in Table 4 and the standard levels in Table 5. In addition to these factors, the experiment was performed under the same welding conditions, and all parts that entered welding were from one stamping batch (for a specific value x_3).

Table 4. Levels of observed factors.

Coded Scale	Natural Scale	Factor Level		
		−1	0	1
x_1	Current—I (A)	160	180	200
x_2	Welding speed—v (cm·min^{-1})	50	60	70
x_3	Stamping size—Z (mm)	315.78	–	316.22

Table 5. Standard levels of observed factors.

Factor	I (A)	v (cm·min^{-1})	Z (mm)
Level	200	70	Variable over series

The experimental design included a total of 8 cube points, 8 axial points, and 10 pseudo-central points. A graphical representation of the experimental design (with five replicates at pseudo-central points) is shown in Figure 11.

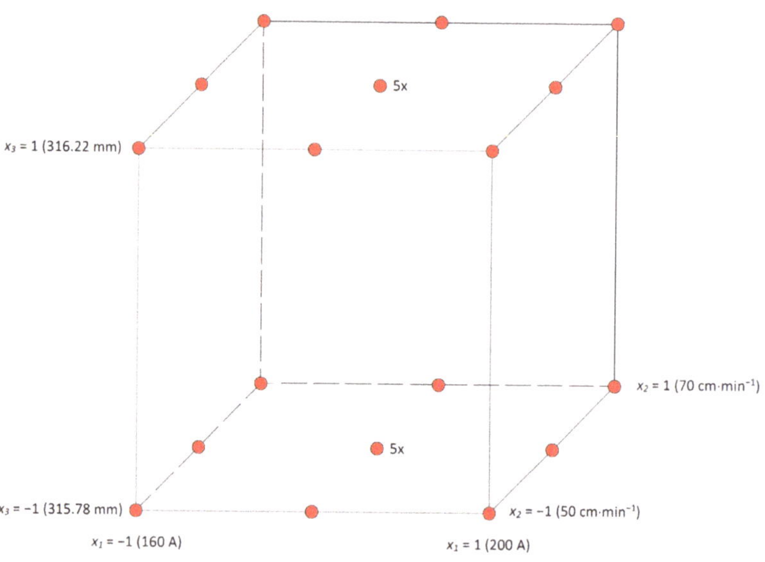

Figure 11. Demonstration of the proposed experiment plan.

In addition to the plan, measurements were performed in order to assess the accuracy of the model for the parameters listed in Table 6.

Table 6. Experiment levels and verification levels.

Factor	Levels Involved in the Model	Levels Not Involved in the Model (Model Verification)
x_1	160, 180, 200	–
x_2	50, 60, 70	52, 54, 56, 58, 62, 64, 66, 68 for all levels of factors x_1 and x_3
x_3	315.78, 316.22	316.08

3. Results and Discussion

When analysing the individual levels of factors and their influence on the resulting length of the welded arm using nonparametric Kruskal–Wallis analysis [24,25] of variance (Table 7), it can be stated that

- at the significance level of 5%, the welding current ((H 2, N = 99) = 6.5254) is a significant factor influencing the change in arm length ($p = 0.0383$),
- at a significance level of 5%, the welding speed ((H 10, N = 99) = 19.7374) is a significant factor influencing the change in arm length ($p = 0.0318$), and
- at the level of significance of 5%, the size of the stamping ((H 2, N = 99) = 70.1071) is a significant factor influencing the change in the length of the arm ($p < 0.001$).

Table 7. Results of Kruskal–Wallis analysis of variance of individual parameters of the experiment.

Current—I (A)	Valid N	Sum of Ranks	Mean Rank
160	33	1950.50	59.1061
180	33	1645.00	49.8485
200	33	1354.50	41.0455

Welding speed—v (cm·min^{-1})	Valid N	Sum of Ranks	Mean Rank
50	9	246.00	27.3333
52	9	286.50	31.8333
54	9	336.00	37.3333
56	9	393.50	43.7222
58	9	435.00	48.3333
60	9	476.00	52.8889
62	9	510.50	56.7222
64	9	530.50	58.9444
66	9	553.00	61.4444
68	9	579.00	64.3333
70	9	604.00	67.1111

Stamping size—Z (mm)	Valid N	Sum of Ranks	Mean Rank
315.78	33	596.00	18.0606
316.08	33	1829.00	55.4242
316.22	33	2525.00	76.5152

Further analysis by multiple comparisons of p values shows that there is a statistically significant difference in the value of the welded arm length at a current value of 160 A and a value of 200 A. Increasing the current value by 40 A causes a decrease in the arm length value by 0.14 mm. The difference between the value of the length of the welded arm at a current of 180 A and 200 A represents a value of 0.07 mm, but this difference is not significant at the significance level of 5%.

Statistically significant differences in the arm length at different speed values are given in Table 8.

Table 8. Statistically significant differences in the arm length at different levels of factor x_2.

Deformation of the Welded Arm (mm)		Increased Factor Level x_2			
		64	66	68	70
The original level of factor x_2	50	0.24	0.26	0.28	0.30
	52	–	0.22	0.24	0.26

Based on the analysis performed by multiple comparisons of p values, it can be further stated that a statistically significant difference was seen in the value of the length of the welded arm at the size of the stamping 315.78 mm and the value 316.22 mm. Increasing the value by 0.44 mm increased the value of the arm length by 0.43 mm. The difference between the value of the length of the welded arm for the 316.08 mm and 316.22 mm stamping is 0.14 mm, and this difference is also significant at the significance level of 5%. The influence of the I and v parameter levels on the mean value of the response Y is shown in Figure 12.

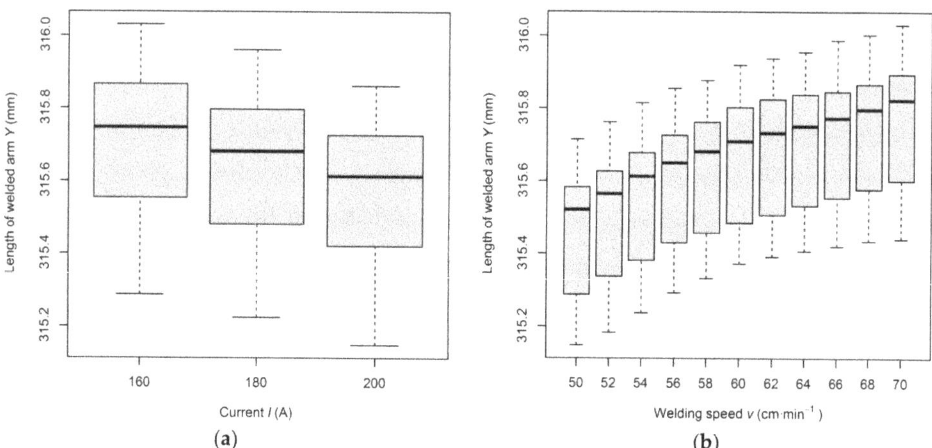

Figure 12. The influence of factor levels on the mean response Y value: (**a**) current I (A); (**b**) welding speed v (cm·min^{-1}).

The degree of overlap, e.g., between the values at the minimum and maximum current level given by the degree of influence of the given factor on the mean value of the response. The overall variability of the response is given by the number of factors and the selection of their levels, and the more significant the influence of the factor on the mean value of the response, the smaller the overlap. From the point of view of the model, this influence is illustrated by the estimation of the model parameters. The smaller the estimated value (influence) of a given factor, the greater the overlap between levels.

Statistical modelling by regression analysis was applied to create a complex dependence of the experimentally investigated welding factors on the value of the final length of the welded component Y. The basic results for the required dependence in a general form can be expressed by Equation (1).

$$Y = f(x_1, x_2, x_3) \tag{1}$$

where x_1—electric current [A], x_2—welding speed [cm·min^{-1}], and x_3—stamping dimension [mm]. The suitability of the used model is documented in Table 9.

Table 9. Suitability of the used model.

Term	Value
RSquare	0.974
RSquare Adj	0.969
Root Mean Square Error	0.044
Mean of Response	315.612
Observations (or Sum Wgts)	26.000

The results show that the predictive power of the model expressed by the adjusted index of determination represents a value of 96.9%. Thus, the model cannot explain 3.1% of the variability of the investigated parameter of the length of the welded arm Y. The average value of the length of the weldment represents a value of 315.612 mm with an average error of 0.044 mm.

From the table of the variance analysis (Table 10), it can be concluded that the variability caused by random errors is significantly less than the variability of the measured values explained by the model. Model F Ratio value implies the model is significant. There is only a 0.01% probability that such a significant F value could occur due to noise.

Table 10. ANOVA table of the model applied.

Source	DF	Sum of Squares	Mean Square	F Ratio	Prob > F
Model	4	1.565	0.391	197.661	<0.0001
Error	21	0.041	0.002	–	–
C. Total	25	1.606	–	–	–

Testing the null (H_0) statistical hypothesis [26,27], which results from the nature of the Fisher–Snedecor test criterion, allows us to conclude that, based on the achieved level of significance $p = 0.0001$, the null hypothesis of the Fisher–Snedecor test criterion can be rejected. It is possible to accept the alternative hypothesis that there is at least one factor whose regression coefficient is statistically different from zero and thus significantly affects the change of the investigated parameter Y. This means that adequate input variables were chosen to describe the change in the dimension of the welded part Y. The model in terms of Fisher–Snedecor test criterion is adequate and significant.

Further testing of the model used was carried out by the so-called lack of fit error test, i.e., the variance of residues and the variance of the measured data within the groups were tested. Thus, it is tested whether the regression model sufficiently captures the observed dependence (Table 11).

Table 11. Insufficient model fit error testing.

Source	DF	Sum of Squares	Mean Square	F Ratio	Prob > F	Max RSq
Lack of Fit	13	0.002	<0.001	0.037	1.000	0.976
Pure Error	8	0.039	0.005	–	–	–
Total Error	21	0.041	–	–	–	–

Given the significance value of 1000 achieved by the Fisher test, a null statistical hypothesis can be accepted for the observed variable Y, which results from the nature of the mismatch error test, and we can say that the model sufficiently captures the variability of experimentally obtained data at the 5% significance level.

Based on the above assumptions and their fulfilment (Tables 10 and 11), the following table (Table 12) presents an estimate of the model parameters with testing the significance of individual effects and their combination at the significance level $\alpha = 0.05$.

Table 12. Estimation of model parameters.

Term	Estimate	Std Error	t Ratio	Prob > |t|	Lower 95%	Upper 95%
Intercept	315.634	0.012	26548.000	<0.0001 *	315.609	315.659
x_1	−0.069	0.013	−5.390	<0.0001 *	−0.096	−0.042
x_2	0.151	0.013	11.740	<0.0001 *	0.124	0.177
$x_2 \cdot x_2$	−0.048	0.018	−2.710	0.013 *	−0.084	−0.011
x_3	0.217	0.009	24.830	<0.0001 *	0.198	0.235

* Significant at the level of significance 5%.

The table for estimating the parameters of the model (Table 12) shows that the welding current on the significance level of 5% has a significant effect on the change in the values of the investigated parameter Y in the range of experimentally used input variables and influence of this input investigated variable on the total value Y variability. Furthermore, it can be concluded that, as the current increases, the conditional arm length value Y also decreases. Another significant input parameter that affects the arm length value is the welding speed, with an overall effect on the variability of the Y value of 32.35%. The influence of the stamping size is also significant, with 55.59% influence on the change in the arm length value Y. It follows from the above that, in the monitored parameter range, the stamping dimension has the most significant influence. In contrast, with other parameters, its influence can be corrected significantly, but not completely.

The output of the model is coded Equation (2)

$$Y = 315.634 - 0.069 \cdot x_1 + 0.151 \cdot x_2 + 0.217 \cdot x_3 - 0.0485119 \cdot x_2^2 \tag{2}$$

The first equation term (intercept) represents the centre of design space. It is clear from the equation that there is a positive correlation between the dimension of the stamping x_3 and the resulting value of the length of the arm Y. Conversely, there is a negative correlation between the welding current x_1 and Y. Thus, the increasing current causes a more significant deformation of the arm, and thus, the resulting arm length Y decreases. The influence of the welding speed x_2 on the resulting value is determined by two members of the equation, while the linear term has a positive correlation and the quadratic term negative correlation. Due to the size of the coefficients of the given members of the equation, the resulting correlation between x_2 and Y is positive. Thus, with increasing welding speed, the total length of the component Y also increases, i.e., the deformation during welding is smaller in comparison. A graphical representation of the resulting influence of individual factors on the value Y is shown in Figure 13.

Based on the above facts, a regression dependence can finally be predicted (3):

$$Y = 2.503 - 3.458 \cdot 10^{-3} \cdot I + 7.209 \cdot 10^{-2} \cdot v + 9.846 \cdot 10^{-1} \cdot Z - 4.750 \cdot 10^{-4} \cdot v^2 \tag{3}$$

This equation is suitable for determining predictions on a scale of selected intervals for individual factors, i.e., for technical practice. At the same time, however, because the intercept is not the centre of design space and the regression coefficients are modified by conversion from coded to actual scale, the equation is not suitable for determining the influence of individual factors for interpreting the model.

Due to the use of pseudo-central points, we performed a model verification near the central level of factor x_3. Verification was performed by analysing the residues of the model for measurements with the stamping size $Z = 316.08$ mm. A graphical representation of the residue analysis is shown in Figure 14 and a summary in Table 13.

It follows from the above that the residues have a normal distribution, which was confirmed by the Shapiro–Wilk test ($p = 0.9979$) and the Kolmogorov–Smirnov test ($p = 0.9406$). We, therefore, verified the correctness of the model in the vicinity of the central level of factor x_3.

The output of the model, the graphic dependence of the length of the welded arm on the change of the current, and the welding speed for the dimension of the stamping $Z = 315.80$ mm are shown in Figure 15.

The model confirms the theoretical assumptions, i.e., that minimal deformations (larger dimension of the welded arm) can be achieved at lower currents and higher welding speeds. The choice of optimal welding parameters can be processed in the form of recommendations for individual dimensions of stampings within the examined interval into a tabular form or graphically in the form of prediction maps. An example for the same dimension of the stamping is shown in Figure 16.

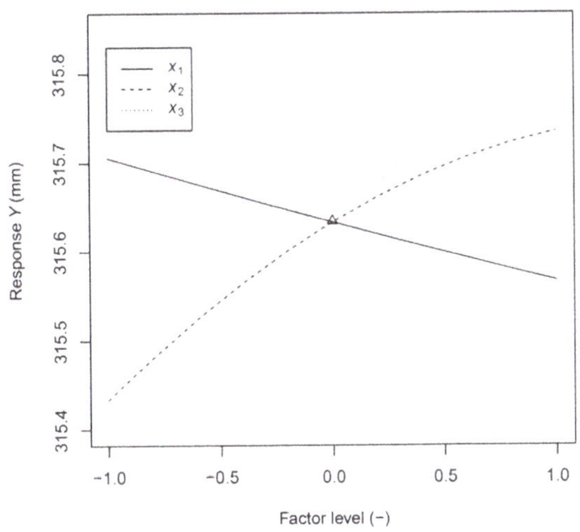

Figure 13. Influence of individual factors on the resulting Response Y.

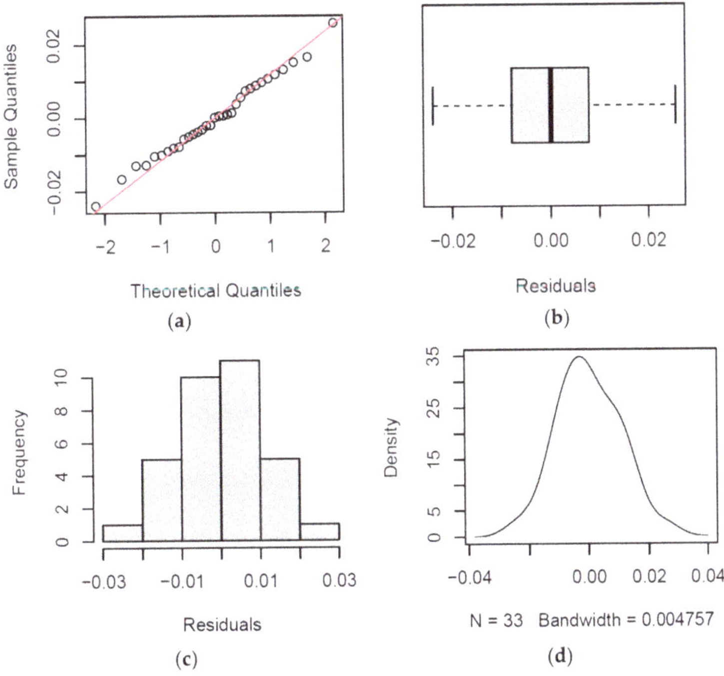

Figure 14. Residue analysis: (**a**) Normal Q–Q Plot; (**b**) Box Plot; (**c**) Histogram; (**d**) Density.

Table 13. Summary of residuals.

Term	Value
Minimum	−0.024
1st quartile	−0.008
Median	0
Mean	0
3rd quartile	0.008
Maximum	0.026
RSquare	0.991
RSquare Adj	0.991

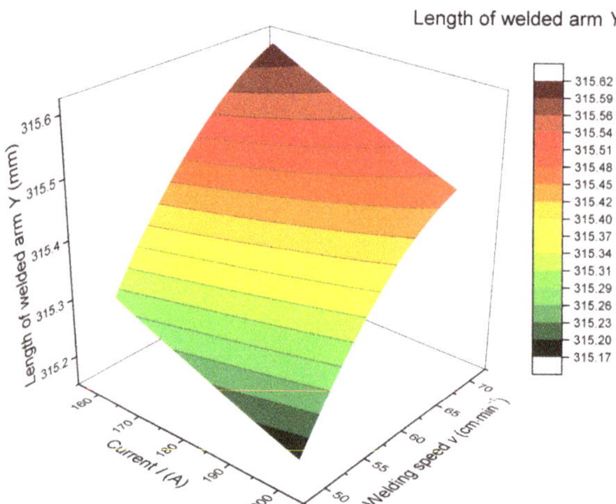

Figure 15. Dependence of the welded arm length on the current and the welding speed for the dimension of the stamping Z = 315.80 mm.

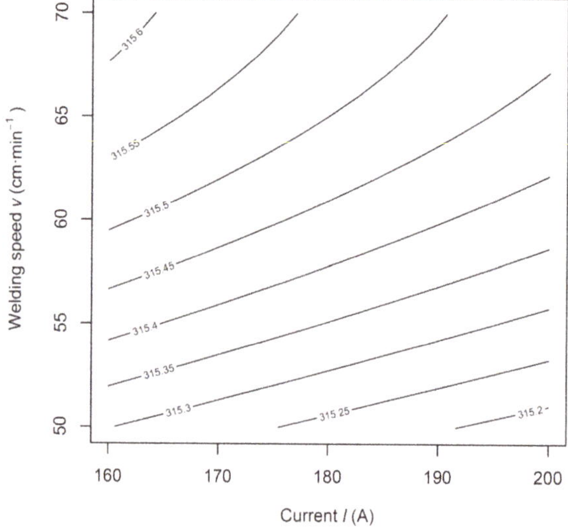

Figure 16. The dependence of the welded arm resulting length on the current and the welding speed for the dimension of the stamping Z = 315.80 mm.

4. Conclusions

This article deals with a significant problem typical of the automotive industry—the dimensional accuracy of welded stampings. We have used the example of a specific component to introduce a new procedure, the result of which is a model that allows the use of deformation during welding to eliminate inaccuracies arising during the stamping process partially. The results can be summarized as follows:

- Methodology that has not been used in this area so far, i.e., the use of pseudo-central points and face-centred axial points in CCD under the assumption of a linear influence of the factor x_3 on the response in the subsequent statistical analysis and modelling allowed us to create a model with a high value of the adjusted coefficient of determination.
- The use of the proposed model allowed us to increase the accuracy of the production or maximise the production while maintaining the required dimensions.
- The influence of basic process parameters, namely welding current and welding speed combined with the changing size of the stamping, was verified.
- The observed linear dependence of thermal deformation on the welding current in combination with a significant curvature of the effect of welding speed (Figure 13) illustrated the complexity of the problem from a physical point of view, as the heat introduced was energy of dissipative nature, which depended not only on the magnitude of the current flowing through and on the time for which the heating process takes place but also on the resistance that the conductive material puts.
- As the indirect measurement of the generated heat for the purpose of precise continuous control is very difficult even with the use of precise measuring instruments, simplified models based on heat input (e.g., I/v) should be implemented only where there are no high demands on model accuracy.
- The unambiguous result of the presented article confirms the theoretical, qualitative assumption that the selected operating parameters have a direct and measurably significant effect on the resulting deformation of the part.
- A specific benefit of the article is the coded Equation (2) with a description and graphical representation of the influence of individual factors (Figure 13) on the final dimension of the welded part.
- The new Equation (3) allows predicting the resulting dimension directly from the entered operating parameters within the considered interval. It is possible to use prediction maps (Figure 16) to set the monitored process optimally in practice.

Author Contributions: Conceptualization, M.K. (Milan Kadnár) and P.K.; methodology, M.K. (Milan Kadnár), M.H., and J.V.; validation, M.G. and J.R.; formal analysis, M.G.; investigation, P.K. and F.T.; data curation, M.B. and M.K. (Milena Kušnerová); writing—original draft preparation, M.H.; writing—review and editing, J.V.; visualization, F.T. and J.R.; project administration, M.H.; funding acquisition, J.R. All authors have read and agreed to the published version of the manuscript.

Funding: This research work has been supported by the grant agency of the Slovak University of Agriculture in Nitra GA 7/19 and VEGA 1/0236/21.

Institutional Review Board Statement: Not applicable.

Informed Consent Statement: Not applicable.

Data Availability Statement: The data presented in this study are available on request from the corresponding author after obtaining permission of authorized person.

Conflicts of Interest: The authors declare no conflict of interest.

References

1. Hong, K.-M.; Shin, Y. Prospects of laser welding technology in the automotive industry: A review. *J. Mater. Process. Technol.* **2017**, *245*, 46–69. [CrossRef]
2. Ogbonna, O.S.; Akinlabi, S.A.; Madushele, N.; Mashinini, P.M.; Abioye, A.A. Application of MIG and TIG Welding in Automobile Industry. *J. Phys. Conf. Ser.* **2019**, *1378*, 042065. [CrossRef]

3. Manladan, S.M.; Abdullahi, I.; Hamza, M.F. A review on the application of resistance spot welding of automotive sheets. *J. Eng. Technol.* **2015**, *10*, 20–37.
4. Muthu, P. Optimization of the Process Parameters of Resistance Spot Welding of AISI 316l Sheets Using Taguchi Method. *Mech. Mech. Eng.* **2019**, *23*, 64–69. [CrossRef]
5. Ma, M.; Lai, R.; Qin, J.; Wang, B.; Liu, H.; Yi, D. Effect of weld reinforcement on tensile and fatigue properties of 5083 aluminum metal inert gas (MIG) welded joint: Experiments and numerical simulations. *Int. J. Fatigue* **2021**, *144*, 106046. [CrossRef]
6. Radaj, D. *Heat Effects of Welding: Temperature Field, Residual Stress, Distortion*; Springer: Berlin/Heidelberg, Germany, 1992; p. 348.
7. Thomas, D.; Galbraith, C.; Bull, M.; Finn, M.J. Prediction of Springback and Final Shape in Stamped Automotive Assemblies: Comparison of Finite Element Predictions and Experiments. *SAE Tech. Pap. Ser.* **2002**. [CrossRef]
8. Önal, A.S.; Kaya, N. Effect and Optimization of Resistance Spot Welding Parameters on the Strength of Welded Hot-Stamped Parts. *Mater. Test.* **2014**, *56*, 466–471. [CrossRef]
9. Hamedi, M.; Shariatpanahi, M.; Mansourzadeh, A. Optimizing spot welding parameters in a sheet metal assembly by neural networks and genetic algorithm. *Proc. Inst. Mech. Eng. Part B J. Eng. Manuf.* **2007**, *221*, 1175–1184. [CrossRef]
10. Perret, W.; Schwenk, C.; Rethmeier, M.; Raphael, T.R.; Alber, U. Case Study for Welding Simulation in the Automotive Industry. *Weld. World* **2011**, *55*, 89–98. [CrossRef]
11. Kim, T.; Park, H. Optimization of welding parameters for resistance spot welding of TRIP steel with response surface methodology. *Int. J. Prod. Res.* **2005**, *43*, 4643–4657. [CrossRef]
12. Caro, P.; Odenberger, E.-L.; Schill, M.; Steffenburg-Nordenström, J.; Niklasson, F.; Oldenburg, M. Prediction of shape distortions during forming and welding of a double-curved strip geometry in alloy 718. *Int. J. Adv. Manuf. Tech.* **2020**, 1–15. [CrossRef]
13. Li, W.; Cheng, S.; Hu, S.J.; Shriver, J. Statistical Investigation on Resistance Spot Welding Quality Using a Two-State, Sliding-Level Experiment. *J. Manuf. Sci. Eng.* **2000**, *123*, 513–520. [CrossRef]
14. Zhang, H.; Hu, S.J.; Senkara, J.; Cheng, S. A Statistical Analysis of Expulsion Limits in Resistance Spot Welding. *J. Manuf. Sci. Eng.* **2000**, *122*, 501–510. [CrossRef]
15. Juang, S.; Tarng, Y. Process parameter selection for optimizing the weld pool geometry in the tungsten inert gas welding of stainless steel. *J. Mater. Process. Technol.* **2002**, *122*, 33–37. [CrossRef]
16. Cho, J. Prediction of Arc Welding Quality through Artificial Neural Network. *J. Weld. Join.* **2013**, *31*, 44–48. [CrossRef]
17. Li, R.; Dong, M.; Gao, H. Prediction of Bead Geometry with Changing Welding Speed Using Artificial Neural Network. *Materials* **2021**, *14*, 1494. [CrossRef] [PubMed]
18. Shiu, B.W.; Shi, J.; Tse, K.H. The dimensional quality of sheet metal assembly with welding-induced internal stress. *Proc. Inst. Mech. Eng. Part D J. Automob. Eng.* **2000**, *214*, 693–704. [CrossRef]
19. Ceglarek, D.; Shi, J. Design Evaluation of Sheet Metal Joints for Dimensional Integrity. *J. Manuf. Sci. Eng.* **1998**, *120*, 452–460. [CrossRef]
20. Hu, S.J.; Wu, S. Identifying sources of variation in automobile body assembly using principal component analysis. *NAMRI/SME* **1992**, *20*, 311–316.
21. Liu, S.C.; Hu, S.J. Sheet Metal Joint Configurations and Their Variation Characteristics. *J. Manuf. Sci. Eng.* **1998**, *120*, 461–467. [CrossRef]
22. Shiu, B.; Ceglarek, D.; Shi, J. Multi-stations sheet metal assembly modeling and diagnostics. *NAMRC* **1996**, *24*, 199–204.
23. Shiu, B.; Apley, D.; Ceglarek, D.; Shi, J. Tolerance allocation for sheet metal assembly using beam-based model. *Trans. IIE* **2003**, *35*, 329–342. [CrossRef]
24. Kruskal, W.H.; Wallis, W.A. Use of ranks in one-criterion variance analysis. *J. Am. Stat. Assoc.* **1952**, *47*, 583–621. [CrossRef]
25. Ostertagová, E.; Ostertag, O.; Kováč, J. Methodology and Application of the Kruskal-Wallis Test. *Appl. Mech. Mater.* **2014**, *611*, 115–120. [CrossRef]
26. Ross, S.M. Testing Statistical Hypotheses. In *Introductory Statistics*; Academic Press: Boston, MA, USA, 2010; pp. 387–442.
27. Sheskin, D.J. *Handbook of Parametric and Nonparametric Statistical Procedures*; Chapman & Hall/CRC: London, UK, 2007.